The Therapeutic Potential of Gold Complexes

The Therapeutic Potential of Gold Complexes

By

Waseem A. Wani
Sri Pratap College, India
Email: waseemorg@gmail.com

Mohamed F. Alajmi
King Saud University, Saudi Arabia
Email: malajmii@ksu.edu.sa

Afzal Hussain
King Saud University, Saudi Arabia
Email: afihussain@ksu.edu.sa

and

Irshad Ahmad Wani
Government Degree College Anantnag, India
Email: irshadwani101@gmail.com

ROYAL SOCIETY OF CHEMISTRY

Print ISBN: 978-183767-888-4
PDF ISBN: 978-183767-889-1
EPUB ISBN: 978-1-83767-581-4

A catalogue record for this book is available from the British Library

The Royal Society of Chemistry is a charity, registered in England and Wales, Number 207890, and a company incorporated in England by Royal Charter (Registered No. RC000524), registered office: Burlington House, Piccadilly, London W1J 0BA, UK, Telephone: +44 (0) 20 7437 8656.

For further information see our website at www.rsc.org

For general enquiries, please contact books@rsc.org

For EU product safety enquiries, please email books@rsc.org or contact Royal Society of Chemistry Worldwide (Germany) GmbH, Römischer Hof, Unter den Linden 10, 10117 Berlin.

Printed in the United Kingdom by CPI Group (UK) Ltd, Croydon, CR0 4YY, UK

Dedicated to Our Parents and Teachers

The Therapeutic Potential of Gold Complexes
By Waseem A. Wani, Mohamed F. Alajmi, Afzal Hussain and Irshad Ahmad Wani

Published by the Royal Society of Chemistry, www.rsc.org

Preface

Metals and metal ions have been used in medicine for more than 5000 years, with a large array of metal atom/ion-containing compounds being used for various medicinal purposes. Advances in research over the past five decades have shown a significant increase in the use of metal-containing compounds for the treatment and management of a plethora of diseases, such as cancer, diseases of bacterial, fungal, viral and parasitic origins, and other inflammatory diseases. The biological activity of metal complexes is often attributed to their binding affinities with important biomolecules, such as enzymes, proteins, nucleic acids and hormones, among others. The fascinating structures of metal complexes offer versatile platforms for the design of drugs, and thus occupy a ground-breaking niche in Inorganic Medicinal Chemistry and Biochemistry.

In this book, we have attempted to highlight and review the advances that have been made in the design and development of gold complexes as therapeutic agents. The main focus of the book is discussion of the anticancer, antibacterial, antifungal, antiviral, antiparasitic and anti-inflammatory properties of diversely structured gold(I) and gold(III) complexes. Different classes of anticancer complexes based on ligands containing sulfur donors, N-heterocyclic carbenes and thiosemicarbazones are discussed. In addition, anticancer cyclometalated gold complexes, multinuclear gold complexes and bioactive ligands containing gold complexes are described. Additionally, advances in the rational design and development of gold(I) and gold(III) complexes containing phosphine-type ligands, N-heterocyclic carbenes, bis(thiosemicarbazonate) ligands, triphenylphosphines, thiosemicarbazones, 9-deazahypoxanthine ligands, substituted purine derivatives, benzyladenine derivatives and cyclometalated phosphine sulfide ligands with antibacterial, antifungal, antiviral, antiparasitic and anti-inflammatory activities are reviewed. Importantly, the mechanisms of the therapeutic activities

The Therapeutic Potential of Gold Complexes
By Waseem A. Wani, Mohamed F. Alajmi, Afzal Hussain and Irshad Ahmad Wani

Published by the Royal Society of Chemistry, www.rsc.org

of these complexes are discussed and the nanoformulations of anticancer gold complexes have been highlighted. Moreover, the toxicity issues encountered for of some gold complexes are also reviewed. Finally, challenges and future perspectives for research pertaining to the development of therapeutic gold complexes is discussed.

Waseem A. Wani
Afzal Hussain
Mohamed F. Alajmi
Irshad Ahmad Wani

About the Authors

Dr Waseem A. Wani is an Assistant Professor at the Department of Chemistry in Sri Pratap College, M.A. Road, Srinagar, Kashmir, J&K, India. Dr Waseem earned his PhD in Chemistry from Jamia Millia Islamia (Central University), New Delhi, in 2014. He has worked as a Postdoctoral Research Fellow at Universiti Teknologi Malaysia (UTM), Malaysia, Institute of Nano Science and Technology (INST), India, and University of Minnesota (UMN), Twin Cities Campus, Saint Paul-Minneapolis, USA. He has research interests in anticancer metallodrugs, nanofunctionalization of anticancer metallodrugs, drug delivery, and bioproducts. He has authored and co-authored more than 70 research publications (original research papers, reviews, editorials, book chapters, books, brief books, and conference papers) of national and international repute. Dr Waseem figures in the Top 2% list of global scientists 2024 released by Stanford University, USA. Dr Waseem is an active reviewer for many reputed journals.

Dr Afzal Hussain is Associate Professor in the Department of Pharmacognosy, College of Pharmacy, King Saud University, Riyadh, Saudi Arabia. He completed his PhD in Chemistry at Jamia Millia Islamia, New Delhi, India, in 2010. He has worked as a Postdoctoral Research Fellow at Sultan Qaboos University, Muscat, Oman, and Aligarh Muslim University, Aligarh, India. Having been admitted as a Member of the Royal Society of Chemistry (MRSC) in 2018, he has 12 years of experience in teaching and research. His areas of research include separation sciences, bioanalysis, drug analysis, cosmeceutical analysis, nutraceutical analysis, and synthesis of novel nanomaterials as drug delivery agents. Dr Hussain completed a research project as Co-investigator entitled "Synthesis and Characterization of Novel Metal-Based Anticancer Drugs" Grant No.: 36-227 funded by KACST, Riyadh. Dr Hussain has published more than 100 papers in international journals and conference proceedings and is an active reviewer for many prominent journals.

The Therapeutic Potential of Gold Complexes
By Waseem A. Wani, Mohamed F. Alajmi, Afzal Hussain and Irshad Ahmad Wani

Published by the Royal Society of Chemistry, www.rsc.org

Professor Mohamed F. Alajmi is Head of the Skills Development Unit at the Department of Pharmacognosy, College of Pharmacy, King Saud University, Riyadh, Saudi Arabia. He completed his PhD in Pharmacy in 2007. Dr Alajmi has published more than 150 papers in international journals and is a Member of many professional societies, including the American Society for Mass Spectrometry and American Society of Pharmacognosy. He has received two prestigious awards from King Saud University: the Excellence Award in Skills Development (2013) and the Creative Learning and Teaching Award (2016).

Dr Irshad Ahmad Wani is an Assistant Professor in the Postgraduate Department of Chemistry, Govt. Degree College Anantnag, Kashmir, J&K, India. Dr Irshad earned his Masters degree in Chemistry in 2007 and subsequently achieved a PhD in Nanochemistry in 2012 from Jamia Millia Islamia, New Delhi, India. Dr Wani boasts an impressive publication record in esteemed international journals, showcasing his prowess in the realm of academia. Furthermore, he has contributed substantially as an author to several book chapters and has served as a co-editor for several edited books, highlighting his multifaceted involvement in scholarly publications. His role as the Topical Editor of S-Cube: Sustainable Solutions and Society (SPAST) Journal underscores his commitment to sustainable practices and societal advancement. In recognition of his outstanding contributions, Dr Wani was honoured with the ICTSGS Global Services Award by the ICTSGS Foundation in 2021 and the B. N. Kailoo Memorial Medal for best presentation by the ISCAS Foundation in 2009.

Contents

The Therapeutic Potential of Gold Complexes
By Waseem A. Wani, Mohamed F. Alajmi, Afzal Hussain and Irshad Ahmad Wani

Published by the Royal Society of Chemistry, www.rsc.org

CHAPTER 1

Introduction

Since time immemorial humans have suffered massively and lives have been lost as a result of different types of diseases, such as cancer, bacterial, fungal, viral and parasitic diseases, and diabetes and inflammatory conditions. The havoc created by the increasing incidence of cancer, emerging and resistant pathogenic strains, amplified trends in diabetes and increasing instances of inflammatory conditions have further compounded the misery of humans.

Cancer is a genetic disease and is known to progress through a process called carcinogenesis which involves multiple steps. Initially, cancer remains confined to a particular body part. However, with time, it spreads to other parts of the body through a process called metastasis. Metastasis is one of the main factors that complicates the treatment of cancer. Currently, cancer is one of the main causes of mortality worldwide. Annually, more than ten million new cancer cases are reported globally.[1] Aside from cardiovascular disease, cancer is the second most common lethal illness.[2] In the United States, cancer is responsible for one in every four deaths. It is the second cause of mortality in the United States overall, and the leading cause of death among people who are younger than 85 years of age. Projections by Siegel and co-workers published by the American Cancer Society indicate that 2 001 140 new cancer cases and 611 720 deaths will be reported in the United States by the end of 2024.[3] In Europe, cancer is the second most common cause of death after cardiovascular disease and Europe has seen a huge loss of lives due to cancer owing to increasing reports of cancer-related deaths. Santucci *et al.* predicted 1 270 800 cancer-related deaths in Europe for the year 2024.[4] In the developing nations, the incidence of cancer cases is expected to increase dramatically to around 70% over the next decade.[5] The prevalence of cancer is increasing alarmingly in India, with approximately 1 461 427 cancer cases reported in India by the end of 2022. The prevalence of

The Therapeutic Potential of Gold Complexes
By Waseem A. Wani, Mohamed F. Alajmi, Afzal Hussain and Irshad Ahmad Wani

Published by the Royal Society of Chemistry, www.rsc.org

cancer in India is such that one in every nine individuals is likely to contract cancer in their lifetime.[6]

Huge quantities of different types of pathogenic and non-pathogenic bacteria exist within the gut and on the skin of humans. A large proportion of these bacteria are known to live in harmony with humans, and also help in boosting immunity, and are thus beneficial for human health.[7] However, there are also a large number of pathogenic bacterial strains that cause infectious diseases in humans, such as tuberculosis, syphilis, anthrax, cholera, tetanus, leprosy and bubonic plague. The most common lethal diseases of bacterial origin include infections of the respiratory system. Globally, around 7.7 million deaths occurred due to different types of bacterial infections in 2019.[8]

Pathogenic fungi have posed significant threats to public health for a long time. A frightening upsurge in invasive fungal infections in hospitalized patients has been recorded during the last couple of years.[9] Immunocompromised patients are often at the receiving end of invasive fungal infections. The combined effects of global warming, frequent international travel and antifungal drug resistance have, upsettingly, increased the incidence and global range of fungal diseases. Therefore, the risks of the prevalence of invasive fungal infections are also growing in the general population.[10] Fungi cause around 13 million infections and 1.5 million deaths annually, and the trend in infections and deaths keeps on increasing annually.[11]

Pathogenic viruses cause a plethora of infectious diseases, such as the common cold, chickenpox, influenza and cold sores. In addition, some serious diseases, such as avian influenza, Ebola virus disease, rabies, severe acute respiratory syndrome (SARS), acquired immuno-deficiency disease (AIDS) and hepatitis B and C, are also caused by viruses. Moreover, viruses are thought to be the causative agents for several diseases that are under investigation.[12] Globally, millions of deaths are caused annually due to different types of viral infection.[13]

In tropical developing countries, parasitic diseases are the main public health issue. Parasitic infections account for the high mortality and low socio-economic development in these countries. Globally, parasitic diseases cause approximately 2.5 million deaths annually.[14] Malaria, hookworm and amoebiasis infect around 1 000 000–2 000 000, 50 000–60 000 and 40 000–110 000 patients annually, respectively.[15] Approximately, 249 million cases of malaria were reported in 85 malaria-endemic countries in 2022.[16]

The prevalence and incidence of immune-mediated inflammatory diseases (IMIDs) varies substantially across different regions of the world. Rheumatoid arthritis, asthma, multiple sclerosis, atopic dermatitis, psoriasis and inflammatory bowel disease accounted for 1.59%, 54.71%, 0.09%, 36.17%, 6.84% and 0.60% of all new IMID cases, respectively, in 2019. Significant regional and global variation in the age-standardized incidence rate (ASR) of IMIDs is observed. The highest ASR was reported in high socio-demographic index (SDI) regions, high-income North America and the USA. The age distribution of incident cases from the six IMIDs differs throughout the human life span.

Worldwide, from 1990 to 2019, the incident cases of IMIDs increased, with an average annual percent change (AAPC) of 0.68, and the ASR decreased, with an AAPC of −0.34. The incident cases increased across six IMIDs. It is important to note that the ASR of rheumatoid arthritis increased, whereas the ASR of asthma, inflammatory bowel disease, multiple sclerosis, psoriasis and atopic dermatitis decreased.[17]

In a nutshell, humans suffer massively due to different types of pathogenic and non-pathogenic diseases, and other immune system disorders. Cancer, bacterial, fungal, viral and parasitic diseases, and diabetes and certain inflammatory conditions, are responsible for millions of deaths throughout the world annually. Thus, it is imperative that effective treatment modalities for the control and management of these above-mentioned diseases and disorders are developed and so researchers the world over are working assiduously to design and develop drugs against these diseases and disorders.

Metals and metal ions have been used in medicine for around 5000 years.[18] A wide variety of metal atom/ion-containing compounds have been used for various medicinal purposes.[19] Advances in research over the past five decades have shown a significant rise in the use of metal-containing compounds for the treatment and management of a plethora of diseases, such as cancer, diseases of bacterial, fungal, viral and parasitic origins, diabetes and other inflammatory diseases.[20–25] The bioactivity of metal complexes is often ascribed to their binding affinity with important biomolecules, including enzymes, proteins, nucleic acids and hormones, among others.[26] The fascinating structures of metal complexes offer multipurpose platforms for the design of drugs, and thus deliver a ground-breaking force in Inorganic Medicinal Chemistry and Biochemistry. The kinetics (rates of ligand exchange) and thermodynamics (metal–ligand bond energies, redox potentials, *etc.*) of metal complexes can be altered by changing the central metal ion/atom, the structures of the ligands and the oxidation states/numbers of the metal ions/atoms.[27,28] These vital properties of metal complexes have prompted researchers to design and develop metallodrugs (metal-based drugs) with exciting therapeutic properties and applications.

The unforeseen revelation of the cytotoxic properties of cisplatin (Figure 1.1) raised the curiosity of researchers for strategic progress in using metal complexes as anticancer agents. The establishment of cisplatin as a cytotoxic agent was followed by the programmed and planned design and development of second- and third-generation analogues of cisplatin (Figure 1.1) and the subsequent design and development of a diverse range of platinum complexes as anticancer agents. Despite the significant anticancer activities of platinum complexes, certain disadvantages linked with platinum metal complexes have led to a paradigm shift in the interest of researchers towards the exploration of metallodrugs using non-platinum metal ions/atoms. This has led to the design and development of a large reservoir of metal complexes using non-platinum metals, such as ruthenium, copper, gold, iron, scandium, titanium, vanadium, chromium, manganese, cobalt, nickel, zinc, gallium, yttrium, zirconium, nobelium, molybdenum, technetium, rhodium,

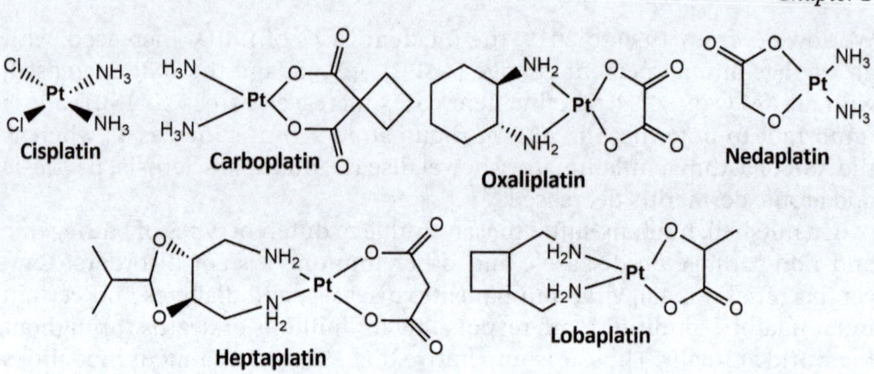

Figure 1.1 Chemical structures of cisplatin, its second-generation analogues (carboplatin, oxaliplatin and nedaplatin) and third-generation analogues (heptaplatin and lobaplatin).

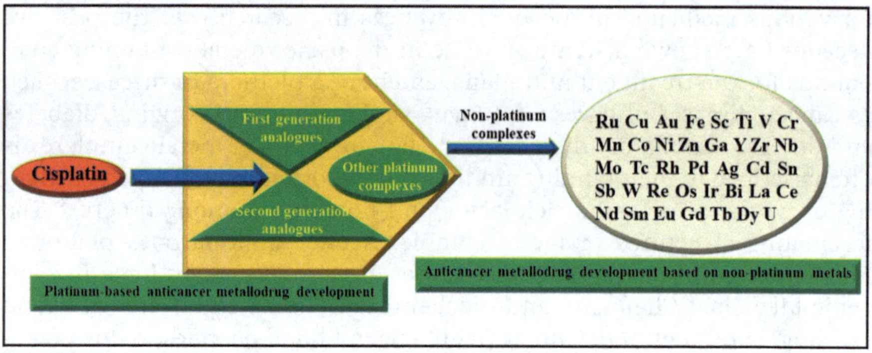

Figure 1.2 A diagrammatic representation of the involvement of platinum and non-platinum metals in anticancer drug development.

palladium, silver, cadmium, tin, antimony, tungsten, rhenium, osmium, iridium, bismuth, lanthanum, cerium, neodymium, samarium, europium, gadolinium, terbium, dysprosium and uranium (Figure 1.2). It is important to note that some metal complexes based on non-platinum metals show exciting activities towards cancer cells and appreciable selectivity towards normal cells. It is worth mentioning that several non-platinum metal complexes are currently being used in the treatment of different cancers, while several complexes are being investigated in clinical trials.[29]

The literature describes the use of several metal ions and metal complexes for the treatment of a variety of bacterial infections.[30] An arsenic-based complex, Salvarsan (arsphenamine or compound 606, Figure 1.3) finds use as an effective drug for treating syphilis.[31] Bismuth and silver-based compounds are being used in different clinical interventions. Infections caused by *Helicobacter pylori* are effectively treated by the use of bismuth co-administered with other antibiotics.[32] Xeroform (Figure 1.3) is a bismuth-based complex

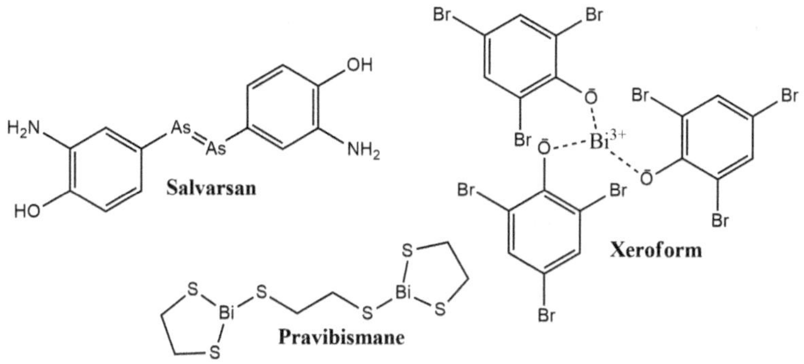

Figure 1.3 Chemical structures of Salvarsan, Xeroform and Pravibismane.

that is used in wound dressing applications.[33] Pravibismane (Figure 1.3) is a bismuth-thiol compound, which completed phase 1b clinical trials for the effective treatment of infected diabetic foot ulcers.[34] Silver has been used for treating and preventing many diseases and infections, since approximately 4000 BCE. The medicinal uses of silver were reported in the literature throughout the 17th and 18th centuries and the antibacterial properties of silver are well documented. Throughout the 19th century, silver nitrate has been used topically to treat ulcerations, burns and infected wounds.[35] In view of the inimitable action mechanisms of metal-containing compounds/complexes in the control, management and treatment of infections of bacterial origin, their single use or synergistic combination with antibiotics provides huge potential for safe and effective treatment approaches against different bacterial infections.

Investigations on the antifungal activities of metal complexes are still rare.[36–38] The lack of proper and adequate mechanistic studies of the antifungal actions of metal complexes can be attributed to this scarcity of available reports. Copper, silver and manganese complexes of phenanthroline did not display toxicity in an *in vivo* model of *Galleria mellonella* (moth larvae) at 10 μg per larva. Interestingly, the fungal burden of larvae was reduced by most of the complexes infected with *Candida haemulonii*.[39,40] Silver complexes of 1,10-phenanthroline-5,6-dione stopped the infecting ability of *Phialophora verrucose* towards *G. mellonella* larvae.[41] Frei *et al.*[42] reported several cobalt(III) complexes of Schiff-base ligands that exhibited appreciable *in vitro* antifungal properties against several fungal strains. In addition, the complexes were non-toxic to the *G. mellonella in vivo* model up to doses of 266 mg kg^{-1}. Rubbiani and co-workers[43] prepared a ferrocene-containing fluconazole complex, which displayed exciting *in vitro* antifungal action in an *in vivo Candida* infection model. The complex decreased the fungal burden as well as enhancing the inflammatory pathology of the kidney and colon in the infected mouse model. In addition to the above studies, a number of metal complexes of different transition metals have shown appreciable antifungal activities against different pathogenic fungi strains. These encouraging reports strongly encourage more investigations of metal complexes for developing antifungal agents.

Antiviral properties of metal complexes are known to depend on the nature of the metal atoms/ions and the types of functional groups in the ligand structures. The antiviral properties of organic antiviral drugs are augmented by formation of their complexes with metal ions/atoms.[44] More than a few antiviral drugs have exhibited improved drug profiles (both pharmacological and toxicological) upon administration in complexed form with metal ions/atoms. Metal complexes of some of the important antiviral drugs, such as ribavirin, acyl-hydrazones, valacyclovir, 2-hydroxybenzamide-based compounds, biguanide derivatives, adefovir and cidofovir have been investigated for the treatment of several viral diseases.[22] Golcu and co-workers, in 2006,[45] documented improved antiviral activities of valacyclovir copper(II) complexes against herpes simplex virus (HSV). It is noteworthy that a number of previous studies have demonstrated promising antiviral properties of metal complexes against several viruses, such as influenza, Ebola, human immunodeficiency virus (HIV) and SARS.[46–51] Overall analysis of the literature on antiviral metal complexes indicates that several metal complexes have shown appreciable antiviral properties against different strains of pathogenic viruses, which strongly suggests extension of further research on the development of antiviral metal complexes.

Bioactive organic drugs in their complexed forms have also been explored for their antiparasitic properties. Several ruthenium, rhodium, copper, platinum and gold complexes of clotrimazole and ketoconazole have been reported and Sanchez-Delgado and co-workers[52] reported the efficacy of ruthenium–clotrimazole complex against *Trypanosoma cruzi*. It is interesting to note that the growth and proliferation of the epimastigotes form of *T. cruzi* was significantly inhibited by all the complexes, and the inhibition was better than that for clotrimazole and ketoconazole. Vietes *et al.*[53] reported Pt, Ru and Pd complexes of 5-nitrofuryl-and 5-nitroacroleine-containing thiosemicarbazones. Better *in vitro* activity towards *T. cruzi* epimastigotes was displayed by Pt(II) and Pd(II) complexes, in comparison to the ligands. The interesting results obtained from the above investigations triggered the rational design and development of several metal complexes with diverse structures for the prevention and treatment of parasitic infections. In a nutshell, a number of metal complexes have exhibited interesting antiparasitic properties against a plethora of parasitic strains, which suggest further research towards the design and development of antiparasitic metal complexes.

Metal complexes have scarcely been investigated for the treatment of anti-inflammatory or autoimmune diseases. Auranofin, Solganal (aurothioglucose) and Myochrysine (sodium aurothiomalate) (Figure 1.4) are some important gold complexes that have found uses in the treatment of rheumatoid arthritis, an autoimmune disorder.[54] Complexes of metals with different anti-inflammatory drugs have been explored for the treatment of inflammatory conditions. Graham and Kettle[55] documented enhanced

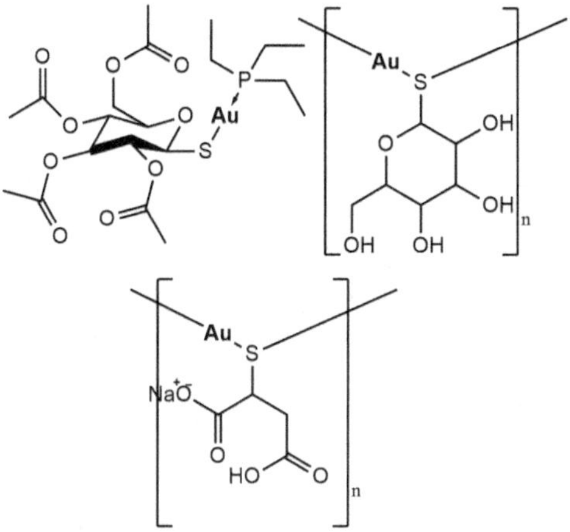

Figure 1.4 Chemical structures of Auranofin, Solganal and Myochrysine.

anti-inflammatory activity of a copper-aspirin complex with reduced side effects in mice, compared to aspirin. Kale *et al.*[56] demonstrated the effect of zinc complexation on the anti-inflammatory action of aceclofenac. The complex and aceclofenac showed similar ability in reducing inflammatory edema in rat paws challenged by carrageenan. Interestingly, the treatment with complex caused fewer ulcers in the stomachs of rats, in comparison with aceclofenac. Narayanachar *et al.*[57] demonstrated the anti-inflammatory action of Co(II), Ni(II), Cu(II) and Zn(II) complexes of Schiff bases derived from 2-mercapto-3-formyl quinoline/2-hydroxy-3-formyl quinoline with 2,6-diaminopyridine. It is interesting to note that most of the complexes remarkably reduced inflammatory edema in rats affected by carrageenan. All the above studies showed encouraging results for the design and development of several metal complexes for the treatment of inflammatory diseases and conditions. Of the complexes developed, several exhibited exciting anti-inflammatory properties with reduced side effects, which warrants further investigation of metal complexes for the treatment of inflammatory diseases.

An in-depth literature survey on SciFinder showed that about 390 publications (original research papers, review papers, books, book chapters and patents) have appeared on "therapeutic gold complexes". A deeper analysis of the published research disclosed that only a few books and book chapters have appeared in the literature.[58–61] The published books are too old to be considered as reference materials and the book chapter is exclusively devoted to a discussion of anticancer gold complexes. Several relevant reviews discussing the anticancer, antimicrobial and antiplasmodial properties of gold complexes have been documented in the last five years.[62–82] However, these reviews have been exclusively about recent advances in the development of anticancer gold complexes, with little or no focus on the development of gold complexes with other therapeutic properties, such as antibacterial, antifungal, antiviral, antiparasitic, antidiabetic and anti-inflammatory.

In view of this, this book aims to review recent advances in the research on therapeutic gold complexes. Different classes of gold complexes with anticancer properties will be discussed and recent advances in the development of antibacterial, antifungal, antiviral, antiparasitic, antidiabetic and anti-inflammatory gold complexes will be described. Attempts have also been made to discuss the nanoformulations of therapeutically active gold complexes. In addition, the mechanisms of the therapeutic action of gold complexes as anticancer, antibacterial, antifungal, antiviral, antiparasitic, antidiabetic and anti-inflammatory agents is discussed. The toxicity of gold complexes with therapeutic properties is also discussed. Finally, the challenges and future research outlook for the design and development of therapeutic gold complexes is discussed.

In response to the distress caused by cancer, diseases of bacterial, fungal, viral and parasitic origins, diabetes and inflammatory conditions throughout human history, the advent of therapeutic metal complexes with various beneficial biological properties has given hope for the treatment and management of these diseases. Globally, researchers are deeply involved in the development of therapeutically effective metal complexes with activity and safety considerations.

References

1. R. Misra, S. Acharya and S. K. Sahoo, *Drug Discovery Today*, 2010, **15**, 842.
2. I. Ali, W. A. Wani and K. Saleem, *Cancer Ther.*, 2011, **8**, 56–70.
3. R. L. Siegel, A. N. Giaquinto and A. Jemal, *CA: Cancer J. Clin.*, 2024, **74**, 12.
4. C. Santucci, S. Mignozzi, M. Malvezzi, P. Boffetta, G. Collatuzzo, F. Levi, C. La Vecchia and E. Negri, *Ann. Oncol.*, 2024, **35**, 308.
5. M. K. Mallath, D. G. Taylor, R. A. Badwe, G. K. Rath, V. Shanta, C. Pramesh, R. Digumarti, P. Sebastian, B. B. Borthakur and A. Kalwar, *Lancet Oncol.*, 2014, **15**, e205.
6. K. Sathishkumar, M. Chaturvedi, P. Das, S. Stephen and P. Mathur, *Indian J. Med. Res.*, 2022, **156**, 598.
7. J. P. McCutcheon, *Annu. Rev. Cell Dev. Biol.*, 2021, **37**, 115.
8. K. S. Ikuta, L. R. Swetschinski and G. R. Aguilar, *Lancet*, 2022, **400**, 2221.
9. M. Hoenigl, D. Seidel, R. Sprute, C. Cunha, M. Oliverio, G. H. Goldman, A. S. Ibrahim and A. Carvalho, *Nat. Microbiol.*, 2022, **7**, 1127.
10. World Health Organization (WHO) Report, *WHO fungal priority pathogens list to guide research, development and public health action*, 2022.
11. F. Bongomin, S. Gago, R. O. Oladele and D. W. Denning, *J. Fungi*, 2017, **3**, 57.
12. A. L. Komaroff, *J. Clin. Virol.*, 2006, **37**, S39.
13. N. Nathanson, *The Human Toll of Viral Diseases: Past Plagues and Pending Pandemics, Viral Pathogenesis*, Elsevier, 3rd edn, 2016.
14. C. J. Schofield, *Parasitol. Today*, 1985, **1**, 181.
15. D. R. Hopkins, *Am. J. Trop. Med. Hyg.*, 1992, **46**, 626.

16. P. Venkatesan, The 2023 WHO World malaria report, *Lancet Microbe*, 2024, **5**, e214.
17. X. Lin, Y. Xu and X. Pan, *Sci. Rep.*, 2020, **10**, 14790.
18. D. Wu, *EClinicalMedicine*, 2023, **64**, 102193.
19. C. Orvig and M. J. Abrams, *Chem. Rev.*, 1999, **99**, 2201.
20. R. A. Alderden, M. D. Hall and T. W. Hambley, *J. Chem. Educ.*, 2006, **83**, 728.
21. T. Gamberi and M. Hanif, *J. Med. Microbiol.*, 2022, **70**, 001363.
22. A. Frei, A. G. Elliott, A. Kan, H. Dinh, S. Bräse, A. E. Bruce, M. R. Bruce, F. Chen, D. Humaidy, N. Jung, A. P. King, P. G. Lye, H. K. Maliszewska, A. M. Mansour, D. Matiadis, M. P. Muñoz, T.-Y. Pai, S. Pokhrel, P. J. Sadler, M. Sagnou, M. Taylor, J. J. Wilson, D. Woods, J. Zuegg, W. Meyer, A. K. Cain, M. A. Cooper and M. A. T. Blaskovich, *JACS Au*, 2022, **2**, 2277.
23. C. Abate, F. Carnamucio, O. Giuffrè and C. Foti, *Biomolecules*, 2022, **12**, 933.
24. S. P. Fricker, R. M. Mosi, B. R. Cameron, I. Baird, Y. Zhu, V. Anastassov, J. Cox, P. S. Doyle, E. Hansell, G. Lau, J. Langille, M. Olsen, L. Qin, R. Skerlj, R. S. Y. Wong, Z. Santucci and J. H. McKerrow, *J. Inorg. Biochem.*, 2008, **102**, 1839.
25. A. Levina and P. A. Lay, *Dalton Trans.*, 2011, **40**, 11675.
26. C. H. Leung, S. Lin, H. J. Zhong and D. L. Ma, *Chem. Sci.*, 2015, **6**, 871.
27. I. Ali, W. A. Wani, K. Saleem and A. Haque, *Anti-Cancer Agents Med. Chem.*, 2013, **13**, 296.
28. I. Romero-Canelón and P. J. Sadler, *Inorg. Chem.*, 2013, **52**, 12276.
29. W. A. Wani, S. Prashar, S. Shreaz and S. Gómez-Ruiz, *Coord. Chem. Rev.*, 2016, **312**, 67.
30. P. F. Iqbal, M. A. Malik and W. A. Wani, *World J. Pharm. Pharm. Sci.*, 2018, **7**, 654.
31. J. W. Alexander, *Surg. Infect.*, 2009, **10**, 289.
32. G. Gasser, *Chimia*, 2015, **69**, 442.
33. K. D. Mjos and C. Orvig, *Chem. Rev.*, 2014, **114**, 4540.
34. D. J. Barillo, A. R. Barillo, S. Korn, K. Lam and P. S. Attar, *Burns*, 2017, **43**, 1189.
35. A. D. Politano, K. T. Campbell, L. H. Rosenberger and R. G. Sawyer, *Surg. Infect.*, 2013, **14**, 8.
36. A. D. Politano, K. T. Campbell, L. H. Rosenberger and R. G. Sawyer, *Surg. Infect.*, 2013, **14**, 8.
37. S. N. Sovari, N. Radakovic, P. Roch, A. Crochet, A. Pavic and F. Zobi, *Eur. J. Med. Chem.*, 2021, **226**, 113858.
38. H. Parveen, R. A. S. Alatawi, M. A. Alsharif, M. I. Alahmdi, S. Mukhtar, S. A. Khan, S. Hasan and A. U. Khan, *Appl. Organomet. Chem.*, 2018, **32**, e4257.
39. J. A. de Azevedo-França, L. P. Borba-Santos, G. de Almeida Pimentel, C. H. J. Franco, C. Souza, J. de Almeida Celestino, E. F. de Menezes, N. P. Dos Santos, E. G. Vieira and A. M. D. C. Ferreira, *J. Inorg. Biochem.*, 2021, **219**, 111401.

40. R. M. Gandra, P. Mc Carron, M. F. Fernandes, L. S. Ramos, T. P. Mello, A. C. Aor, M. H. Branquinha, M. McCann, M. Devereux and A. L. Santos, *Front. Microbiol.*, 2017, **8**, 1257.
41. R. M. Gandra, P. McCarron, L. Viganor, M. F. Fernandes, K. Kavanagh, M. McCann, M. H. Branquinha, A. L. S. Santos, O. Howe and M. Devereux, *Front. Microbiol.*, 2020, **11**, 1.
42. I. M. Keseler, S. Gama-Castro, A. Mackie, R. Billington, C. Bonavides-Martínez, R. Caspi, A. Kothari, M. Krummenacker, P. E. Midford and L. Muñiz-Rascado, *Front. Microbiol.*, 2021, **12**, 711077.
43. A. Frei, A. P. King, G. J. Lowe, A. K. Cain, F. L. Short, H. Dinh, A. G. Elliott, J. Zuegg, J. J. Wilson and M. A. Blaskovich, *Chem. – Eur. J.*, 2021, **27**, 2021.
44. R. Rubbiani, T. Weil, N. Tocci, L. Mastrobuoni, S. Jeger, M. Moretto, J. Ng, Y. Lin, J. Hess, S. Ferrari, A. Kaech, L. Young, J. Spencer, A. L. Moore, K. Cariou, G. Renga, M. Pariano, L. Romani and G. Gasser, *RSC Chem. Biol.*, 2021, **2**, 1263.
45. A. Nourian and H. Khalili, *Acta Bio Med. Atenei Parmensis*, 2020, **91**, 239.
46. A. Gölcü, M. Dolaz, H. Demirelli, M. Diðrak and S. Serin, *Transition Met. Chem.*, 2006, **31**, 658.
47. I. Yousuf, M. Bashir, F. Arjmand and S. Tabassum, *Coord. Chem. Rev.*, 2021, **445**, 214104.
48. M. Chakravarty and A. Vora, *Drug Delivery Transl. Res.*, 2021, **11**, 748.
49. R. E. F. De Paiva, A. M. Neto, I. A. Santos, A. C. G. Jardim, P. P. Corbi and F. R. G. Bergamini, *Dalton Trans.*, 2020, **49**, 16004.
50. M. Kowalczyk, A. Golonko, R. Świsłocka, M. Kalinowska, M. Parcheta, A. Swiergiel and W. Lewandowski, *Front. Pharmacol*, 2021, **12**, 709104.
51. N. Maldonado and P. Amo-Ochoa, *Dalton Trans.*, 2021, **50**, 2310.
52. G. R. Tortella, J. C. Pieretti, O. Rubilar, M. Fernández-Baldo, A. Benavides Mendoza, M. C. Diez and A. B. Seabra, *Crit. Rev. Biotechnol.*, 2021, **42**, 431.
53. R. A. Sanchez-Delgado, K. Lazardi, L. Rincon, J. A. Urbina, A. J. Hubert and A. N. Noels, *J. Med. Chem.*, 1993, **36**, 2041.
54. M. Vieites, P. Smircich, B. Parajón-Costa, J. Rodríguez, V. Galaz, C. Olea-Azar, L. Otero, G. Aguirre, H. Cerecetto and M. González, *JBIC, J. Biol. Inorg. Chem.*, 2008, **13**, 723.
55. K. H. Thompson and C. Orvig, *J. Inorg. Biochem.*, 2006, **100**, 1925.
56. J. E. Philip, M. Shahid, M. P. Kurup and M. P. Velayudhan, *J. Photochem. Photobiol., B*, 2017, **175**, 178.
57. N. Patel, A. K. Prajapati, R. N. Jadeja, I. P. Tripathi and N. Dwivedi, *J. Indian Chem. Soc.*, 2021, **98**, 100047.
58. Y. Yoshikawa, E. Ueda, H. Miyake, H. Sakurai and Y. Kojima, *Biochem. Biophys. Res. Commun.*, 2001, **281**, 1190.
59. X. Wang, J. Du, T. Zhou, X. Fang and H. Yang, *J. Mol. Struct.*, 2023, **1292**, 136141.
60. C. H. Leung, S. Lin, H. J. Zhong and D. L. Ma, *Chem. Sci.*, 2015, **6**, 871.
61. G. G. Graham and A. J. Kettle, *Biochem. Pharmacol.*, 1998, **1**, 307.
62. M. A. Kale, R. Shelke and R. B. Nawale, *Anti-Inflammatory Anti-Allergy Agents Med. Chem.*, 2014, **13**, 36.

63. S. D. Dhumwad, S. Mutalik, M. H. Hugar and P. N. Naik, *Main Group Chem.*, 2013, **12**, 87.
64. P. J. S. Sadler and E. Rodney, *Bioinorg. Chem.*, 1995, **2**, 1039.
65. A. Casini, R. Sun and I. Ott, *Met. Ions Life Sci.*, 2018, **18**, 9783110470734.
66. M. Y. Wani and M. A. Malik, *Gold and its Complexes in Anticancer Chemotherapy*, Springer Singapore, 2022.
67. B. Kirkham and M. S. Hoy, Undesirable side effects of therapeutic gold complexes, in *Handbook of Metal-Ligand Interactions in Biological Fluids: Bioinorganic Medicine*, ed. G. Berthon, 1995, vol. 2, p. 1413.
68. Y. Ma, X. Chang, X. Liang, Z. Lv, L. Shan, M. Lu, Q. Wen, Z. Gust and R. L. Wukun, *Chem. Soc. Rev.*, 2022, **51**, 5518.
69. S. Michele and B. Benoit, *Eur. J. Inorg. Chem.*, 2023, **26**, e202300340.
70. N. A. Yusoh, H. Ahmad and M. R. Gill, *ChemMedChem*, 2020, **15**, 2121.
71. C. K. Adokoh, *RSC Adv.*, 2020, **10**, 2975.
72. D. Fontinha, S. A. Sousa, T. S. Morais, M. Prudencio, J. H. Leitao, G. Y. Le Lorcy, D. Silva, A. L. R. Velho, F. G. Mariana and D. Belo, *Metallomics*, 2020, **12**, 974.
73. G. Moreno-Alcantar, P. Picchetti and A. Casini, *Angew. Chem., Int. Ed.*, 2023, **62**, e202218000.
74. J. Zhang, Y. Li, R. Fang, W. Wei, Y. Wang, J. Jin, F. Yang and J. Chen, *Front. Pharmacol*, 2022, **13**, 979951.
75. S. Y. Hussaini, R. A. Haque and M. R. Razali, *J. Organomet. Chem.*, 2019, **882**, 96.
76. Z. Yang, G. Jiang, Z. Xu, S. Zhao and W. Liu, *Coord. Chem. Rev.*, 2020, **423**, 213492.
77. Y. Liu, Y. Lu, Z. Xu, X. Ma, X. Chen and W. Liu, *Drug Discovery Today*, 2022, **27**, 1961.
78. J. Ceramella, A. Mariconda, D. Iacopetta, C. Saturnino, A. Barbarossa, A. Caruso, C. Rosano, M. S. Sinicropi and P. Longo, *Bioorg. Med. Chem. Lett.*, 2020, **30**, 126905.
79. N. Mirzadeh, T. S. Reddy and S. K. Bhargava, *Coord. Chem. Rev.*, 2019, **388**, 343.
80. R. T. Mertens, S. Gukathasan, A. S. Arojojoye, C. Olelewe and S. G. Awuah, *Chem. Biomed. Appl. Chem. Rev.*, 2023, **123**, 6612.
81. S. Khan, F. A. Alhumaydhi, M. M. Ibrahim, A. Alqahtani, M. Alshamrani, A. S. Alruwaili, A. A. Hassanian and S. Khan, *Anti-Cancer Agents Med. Chem.*, 2022, **22**, 3086.
82. M. Mora, M. C. Gimeno and R. Visbal, *Chem. Soc. Rev.*, 2019, **48**, 447.

CHAPTER 2

Emergence of Gold and Its Complexes in Medicine

Gold (symbol: Au, atomic number: 79) is a naturally occurring transition element belonging to group 11 of the periodic table. Gold in pure form has a bright and slightly reddish yellow appearance. In terms of its physical properties, gold is soft, malleable and ductile. In terms of chemical reactivity, gold is one of the least chemically reactive elements among the transition metals.[1]

The medicinal use of gold dates back to around 2500 BC, and the earliest application of gold as a therapeutic agent was reported in China.[2] The discovery of the antibacterial potential of gold cyanide, $K[Au(CN)_2]$, against *Tuberculosis bacillus* by Robert Koch in 1890 triggered the use of gold and its compounds in modern medicine. Subsequently, in the 1920s, gold therapy was introduced for the treatment of tuberculosis. During this time, it was also suggested that *Tuberculosis bacillus* caused rheumatoid arthritis, which therefore stimulated the use of gold therapy for rheumatoid arthritis. Unfortunately, the treatment of tuberculosis with gold proved to be ineffective. However, following debate of over 30 years, the Empire Rheumatism Council backed a clinical study that established the efficacy of gold compounds against rheumatoid arthritis. Since then a number of gold-containing drugs have been used for the treatment of several other rheumatic diseases, such as juvenile arthritis, psoriatic arthritis, palindromic rheumatism and discoid lupus erythematosus.[3] Jacques Forestier, in 1929, used injections of gold sodium aurothiomalate (Myochrysine) to treat patients with rheumatoid arthritis. The injections were reported to lessen the pain in joints and also removed edema. However, the treatment with injections was very expensive and had to be carried out for several months in order to achieve the desired results. In addition, gold therapy exhibited some exciting results in the treatment regimens for several skin inflammatory disorders, such as urticaria,

The Therapeutic Potential of Gold Complexes
By Waseem A. Wani, Mohamed F. Alajmi, Afzal Hussain and Irshad Ahmad Wani

Published by the Royal Society of Chemistry, www.rsc.org

pemphigus and psoriasis. Chrysotherapy, which involves treatment of diseases and disorders with gold-based drugs, is a recognized part of modern medicine. Interestingly, patients who underwent chrysotherapy for arthritis had lower evidence of malignancy rates and it was thought that gold compounds might possess cytotoxic effects.[4–8]

A breakthrough development took place in the medicinal history of gold compounds during the 1970s and 1980s. Auranofin, an Au(I) phosphine compound, was developed by Sutton *et al.* for the treatment of rheumatoid arthritis. In 1985, the drug was approved for clinical use.[9–11] Later, auranofin was explored as a treatment option for several diseases, including neurodegenerative disorders, cancer, parasitic infections, bacterial infections and AIDS. Finally, auranofin received United States Food and Drug Administration (US FDA) approval for phase II clinical trials for the treatment of cancer.[12] This development fascinated researchers, encouraging further studies towards the rational design and development of gold-based anticancer drugs. The interest in the investigation of gold complexes as anticancer agents is due to the fact that gold in the +III oxidation state is isoelectronic with the platinum(II) ion. Therefore, gold(III) complexes exhibit square-planar geometries, similar to those exhibited by some highly active platinum(II) anticancer complexes, such as cisplatin. In addition, the mammalian cellular environment is mostly reducing, and so gold(III) complexes can be easily reduced to gold(I) complexes and metallic gold *in vivo*.[13] However, suitable choices of ligand donor atoms serve to stabilize the gold(III) ion in the complexes.

The literature describes the use of colloidal gold for curing infections.[14] Advances in nanotechnology, in combination with medicine, have made it possible to use gold nanoparticles or surface-modified gold nanoparticles in anticancer chemotherapy.[15,16] For the diagnosis and treatment of cancer, gold nanoparticles are the preferred agents owing to their biocompatibility and unique optical properties.[17] Gold nanoparticles easily allow their surface functionalization with important biomolecules, such as DNA, small interfering (si) RNA, antibodies, peptides and receptors, due to the efficient binding affinity of gold for thiol and amine groups. This potential of gold nanoparticles to be functionalized by multiple ligands with remarkable loading capacity makes them promising vehicles for use in drug delivery, labeling, sensing and imaging.[18–20] At present, Aurimmune and AuroShell, which are gold-based nanomaterials, are undergoing clinical trials and await approval by the US FDA.

So far, a large pool of gold compounds with diverse ligand architectures have been reported with promising activities against different cancer lines and with reduced toxicity towards normal cells.[21] In addition, a plethora of gold complexes with remarkable activities against different strains of pathogenic bacteria, fungi, viruses and parasites have been reported.[22] Interesting gold-based complexes with significant antidiabetic activities and potent anti-inflammatory properties have also been reported.[23] Overall, researchers worldwide are keenly working on the rational design and development of therapeutic gold complexes.

References

1. H. Hassan, P. Sharma, M. R. Hasan, S. Singh, D. Thakur and J. Narang, *Mater. Sci. Energy Technol.*, 2022, **5**, 375.
2. H. Zhao and Y. Ning, *Gold Bull.*, 2001, **34**, 24.
3. L. Messori and G. Marcon, Gold complexes in the treatment of rheumatoid arthritis, in *Metal Ions and Their Complexes in Medication*, ed. A. Sigel, CRC Press, 2004, pp. 280–301.
4. E. R. T. Tiekink, *Crit. Rev. Oncol. Hematol.*, 2002, **42**, 225.
5. I. Lorenzen, *Ann. Clin. Res.*, 1975, **7**, 195.
6. J. F. Fries, D. Bloch, P. Spitz and D. M. Mitchell, *Am. J. Med.*, 1985, **78**, 56.
7. J. R. Ward, *Am. J. Med.*, 1988, **85**, 39.
8. L. Maiore, M. A. Cinellu, E. Michelucci, G. Moneti, S. Nobili, I. Landini, E. Mini, A. Guerri, C. Gabbiani and L. Messori, *J. Inorg. Biochem.*, 2011, **105**, 348.
9. W. F. Kean, L. Hart and W. W. Buchanan, Auranofin, *Br. J. Rheumatol.*, 1997, **36**, 560.
10. B. M. Sutton, E. McGusty, D. T. Walz and M. J. DiMartino, *J. Med. Chem.*, 1972, **15**, 1095.
11. M. Sutton, *Gold Bull.*, 1986, **19**, 15.
12. C. Roder and M. J. Thomson, *Drugs R&D*, 2015, **15**, 13.
13. B. Bertrand, M. R. M. Williams and M. Bochmann, *Chem. – Eur. J.*, 2018, **24**, 11840.
14. M. C. Daniel and D. Astruc, *Chem. Rev.*, 2004, **104**, 293.
15. J. D. Gibson, B. P. Khanal and E. R. Zubarev, *J. Am. Chem. Soc.*, 2007, **129**, 11653.
16. G. F. Paciotti, D. G. Kingston and L. Tamarkin, *Drug Dev. Res.*, 2006, **67**, 47.
17. P. Ghosh, G. Han, M. De, C. K. Kim and V. M. Rotello, *Adv. Drug Delivery Rev.*, 2008, **60**, 1307.
18. D. A. Giljohann, D. S. Seferos, W. L. Daniel, M. D. Massich, P. C. Patel and C. A. Mirkin, *Angew. Chem., Int. Ed.*, 2010, **49**, 3280.
19. D. Pissuwan, T. Niidome and M. B. Cortie, *J. Controlled Release*, 2011, **149**, 65.
20. K. E. Sapsford, L. Berti, I. L. Medintz and T. P. Begley, Fluorescence spectroscopy: applications in chemical biology, in *Wiley Encyclopedia of Chemical Biology*, Wiley, 2007.
21. G. Moreno-Alcántar, P. Picchetti and A. Casini, *Angew. Chem., Int. Ed.*, 2023, **62**, e202218000.
22. B. Đ. Glišić and M. I. Djuran, *Dalton Trans.*, 2014, **28**, 5950.
23. J. B. Krajewska, J. Włodarczyk, D. Jacenik, R. Kordek, P. Taciak, R. Szczepaniak and J. Fichna, *Int. J. Mol. Sci.*, 2021, **22**, 3121.

CHAPTER 3

Gold Complexes as Anticancer Agents

3.1 Gold Complexes as Anticancer Agents

Gold complexes have shown enormous potential for the design and development of anticancer drugs.[1–5] Gold complexes have displayed remarkable *in vitro* anticancer activity against cancer cells, including platinum-resistant ones. Since the United States Food and Drug Administration's (US FDA's) approval of aurothiomalate, aurothioglucose and auranofin for treating arthritis,[6–12] the development of therapeutic gold complexes has received considerable interest from researchers worldwide and the anticancer activities and applications of gold complexes of phosphines,[13] dithiocarbamates,[14] porphyrins,[15] N-heterocyclic carbenes (NHCs)[16] and alkynyls, among others,[17] have been documented. At present, the choice of gold as a preferred metal center for the design and development of anticancer agents is due to the fact that gold(III) is isoelectronic with the platinum(II) ion. Therefore, complexes of gold(III) also adopt square-planar geometries, similar to those adopted by cisplatin and its second- and third-generation analogues. Unfortunately, the mammalian cellular environment is mostly reducing, thus gold(III) complexes may easily be reduced to gold(I) complexes and metallic gold *in vivo*.[7] However, suitable choice of ligand donor atoms can serve to stabilize the gold(III) ion in these complexes. Recent advances in the design and development of anticancer gold complexes are described in the following sections.

3.2 Complexes with Sulfur Donor Ligands

Several gold complexes with ligands encompassing only sulfur atoms as donors have been reported, with promising anticancer properties and this section describes recent advances in the design and development of these

The Therapeutic Potential of Gold Complexes
By Waseem A. Wani, Mohamed F. Alajmi, Afzal Hussain and Irshad Ahmad Wani

Published by the Royal Society of Chemistry, www.rsc.org

complexes. Dithiocarbamate ligands have sulfur atoms as donor atoms, and several metal complexes of dithiocarbamate ligands have been studied as anticancer agents.[18] In line with the interest in developing gold complexes of dithiocarbamates as anticancer agents, Altaf *et al.*[19] reported anticancer activities and toxicities of a series of six dialkyl/diaryldithiocarbamato gold(III) complexes (Figure 3.1, complexes **1–6**) against human breast carcinoma (MCF-7), human ovarian carcinoma (A2780) and cisplatin-resistant human ovarian carcinoma (A2780R) cancer cell lines and the normal human fetal lung fibroblast cells (MRC5). Complex **4** was the most active complex, with significantly higher anticancer activity compared to cisplatin against A2780R and MCF-7 cell lines and almost identical activity to that of cisplatin against A2780 cells.

Gold(III) 1,2-dithiolene complexes have been found to display anticancer properties.[20] Sousa and co-workers[21] documented the anticancer activities of four monoanionic gold-bisdithiolate complexes (Figure 3.2, complexes **7–10**) against cisplatin-sensitive and cisplatin-resistant ovarian cancer cells, A2780 and A2780cisR. All the complexes displayed high anticancer activities against both cancer cell lines, with half maximal inhibitory concentration (IC_{50}) values in the range 0.9–5.5 μM upon incubation for 48 h. Complexes **8, 9** and **10** caused the generation of reactive oxygen species (ROS) in a dose-dependent manner. In addition, complexes **7, 8** and **9** activated caspase 3/7, which suggested the involvement of apoptosis in causing the death of the cancer cells. In contrast to cisplatin, complexes **9** and **10** did not damage DNA. This report suggests that monoanionic gold-bisdithiolates, and in particular complex **9**, are promising for further investigation as anticancer drugs. Investigation of the anticancer profiles of golf(III) dithiolenes was further carried out by Fontinha *et al.*[22] The authors

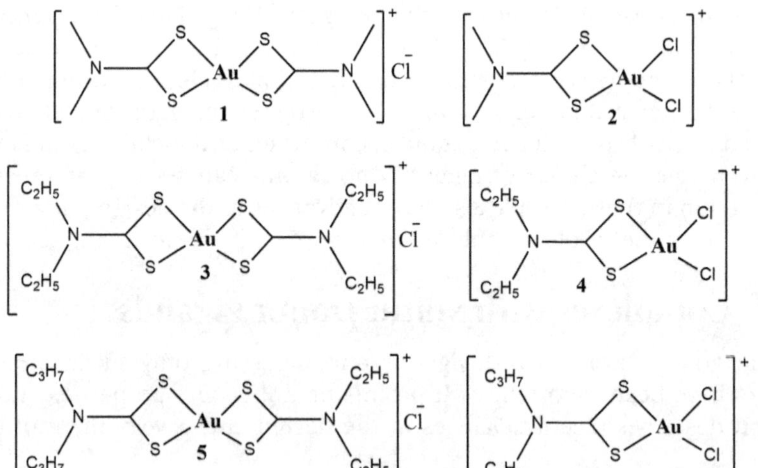

Figure 3.1 Chemical structures of dialkyl/diaryldithiocarbamato gold(III) complexes (**1–6**). Adapted from ref. 19 with permission from The Royal Society of Chemistry.

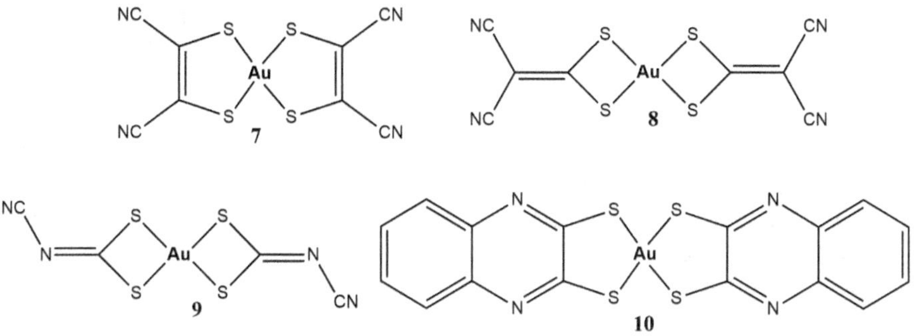

Figure 3.2 Chemical structures of monoanionic gold-bisdithiolate complexes (**7–10**). Adapted from ref. 21 with permission from Elsevier, Copyright 2020.

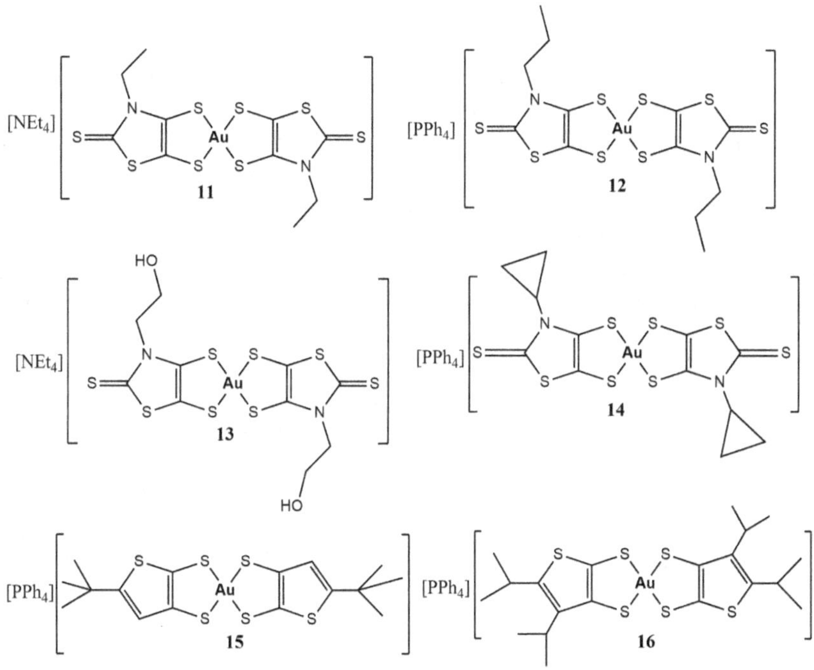

Figure 3.3 Chemical structures of gold(III)-bis(dithiolene) complexes (**11–16**). Adapted from ref. 22 with permission from The Royal Society of Chemistry.

reported the anticancer activities of six gold(III) bis(dithiolene) complexes (Figure 3.3, complexes **11–16**) against A2780/A2780cisR ovarian cancer cell lines. Significant anticancer activities were displayed by complexes **11–16** against A2780/A2780cisR cancer cells, with IC_{50} values in the range 0.08–2 mM. Interestingly, the complexes also overcame cisplatin resistance in A2780cisR cells. It was further revealed, through mechanistic investigations, that thioredoxin reductase (TrxR) is a probable target for the anticancer action of these complexes. In conclusion, this research report indicated a promising future for the development of anticancer complexes based on gold(III) dithiolenes.

In summary, several gold complexes with ligands containing sulfur atoms as donor atoms have been reported with high activity and low toxicity. Several complexes have displayed higher anticancer activity than cisplatin towards cisplatin-sensitive and cisplatin-resistant cancer cell lines. In addition, the mechanisms of action of the complexes are different from that of cisplatin. Consequently, these complexes are promising and should be investigated in *in vivo* models for the development of anticancer metallodrugs.

3.3 Complexes with N-Heterocyclic Carbenes

NHC ligands have been used for the design and development of metal complexes with promising anticancer properties.[23,24] When used as ligands, NHCs exhibit similar bonding properties to those of phosphines.[25–28] In addition, NHC ligands behave as two-electron σ-donors, with a slight π-accepting capacity.[29,30] NHC ligands have several properties that are similar to tertiary phosphines; however, these ligands bind more strongly with the metal atoms/ions and are also brilliant electron donors. In view of the therapeutic utility of NHCs and their metal complexes, Kızrak *et al.*[31] documented the anticancer properties of amine-functionalized gold NHCs against three human cancer cell lines, neuroblastoma (SH-SY5Y), colorectal carcinoma (HTC116) and liver cancer (HEP3B), and the healthy human cell line, alveolar macrophages (HF). The complexes were active against the tested cell lines, with IC_{50} values in the range 4.74 ± 0.91 to 30.98 ± 3.65 μM. Complex **17** (Figure 3.4) showed the maximum anticancer activity against SH-SH5Y (IC_{50} = 4.74 ± 0.91 μM) and HTC116 (IC_{50} = 7.14 ± 1.14 μM). In addition, this complex was non-toxic towards HF cells (54.16 ± 6.85 μM). Complex **18** (Figure 3.4) also displayed appreciable anticancer activity towards HEP3B (6.2 ± 1.12 μM) and SH-SY5Y (8.15 ± 1.24 μM), with selectivity towards the normal cell line. Thus, complexes **17** and **18** may be evaluated against other cancer lines *in vitro*. The

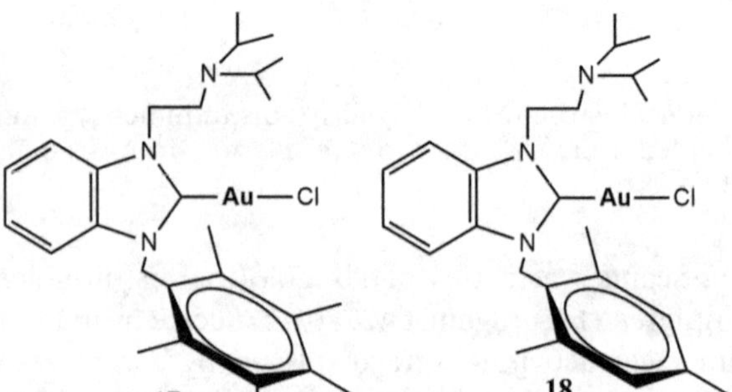

Figure 3.4 Chemical structures of amine-functionalized gold-NHCs (**17** and **18**). Adapted from ref. 31 with permission from Elsevier, Copyright 2019.

reported *in vitro* profiles are quite impressive, which prompted *in vivo* studies of the complexes as well. Sen and co-workers[32] reported anticancer activities of hydroxyl-functionalized heteroleptic bis-NHC-gold(I) complexes. The complexes displayed high anticancer activity against the A549 cell line, with IC_{50} values in the range 0.11–0.56 μM. Interestingly, complex **19** (Figure 3.5), containing doxorubicin, was redirected to the mitochondria, which was surmised to be the reason for its reduced potency compared to doxorubicin. This is thought to eliminate or reduce the nuclear effects that are associated with cardiotoxicity.[33] Thus, the biological profiles of complex **19** and the other complexes reported in this research report should be further evaluated. Büssing *et al.*[34] documented anticancer activities of gold(I) and gold(III) NHCs against non-small cell lung carcinoma (A549) and breast carcinoma (MDA-MB-231) cancer cell lines. Complexes **20–23** (Figure 3.6) were the most

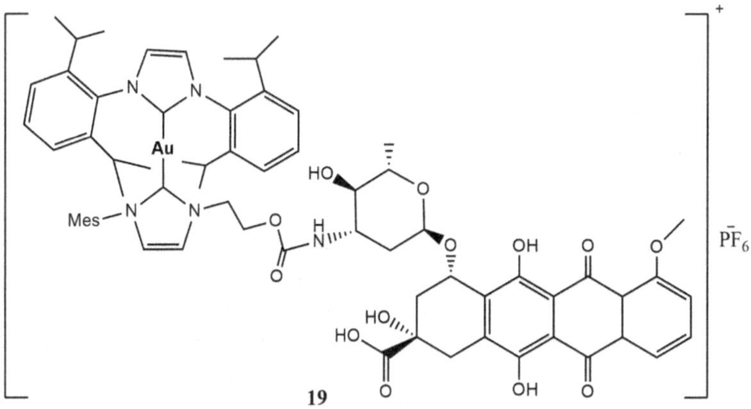

19

Figure 3.5 Chemical structure of doxorubicin-functionalized heteroleptic bis-NHC-gold(I) complex (**19**). Adapted from ref. 32 with permission from The Royal Society of Chemistry.

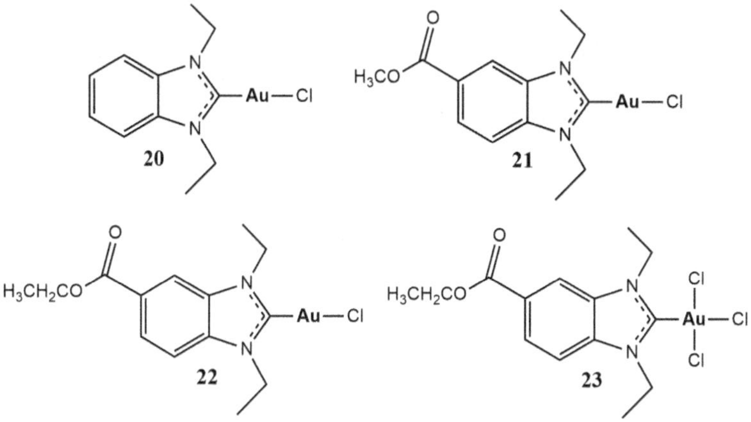

20 **21** **22** **23**

Figure 3.6 Chemical structures of gold(I) and gold(III) NHCs (**20–23**). Adapted from ref. 34 with permission from John Wiley and Sons, Copyright 2021.

active compounds against the tested cell lines. However, all the complexes were less active than the standard drugs, K[Au(CN)$_2$] and auranofin.

In summary, this section shows that several gold complexes with NHCs have been reported that possess remarkable anticancer activity towards cancer cells and good selectivity towards normal cells. Several complexes have shown anticancer effects towards cisplatin-resistant cancer cell lines. In addition, some complexes have redirected and changed the mechanisms of action of certain established anticancer drugs, thereby reducing their toxic effects. In view of the findings described in this section, gold-NHC complexes look quite promising and should be investigated further for the development of safe and effective anticancer metallodrugs.

3.4 Complexes with Thiosemicarbazones

Thiosemicarbazones (TSCs) have always fascinated researchers owing to their range of therapeutic effects. TSCs have been extensively designed and developed for their anticancer effects. Several TSCs have shown activities against various types of cancers, including pancreatic cancer, leukemia, non-small cell lung cancer, breast cancer, prostate cancer, cervical cancer and bladder cancer. In line with the therapeutic benefits of using TSCs in drug development rationales, Zhang *et al.*[35] reported the anticancer activities of a series of gold(III) α-N-heterocyclic thiosemicarbazone complexes against gastric cancer cells. Complex **24** (Figure 3.7) was the most active agent, with substantial activity. Importantly, the authors attempted to synergistically combine the effects of chemotherapy and immunotherapy by causing polarization of tumor-associated macrophages and induction of apoptosis in gastric cancer cells. Possato and co-workers[36] documented the anticancer properties and affinities towards biological targets of two Au(III) thiosemicarbazonate complexes (Figure 3.8, complexes **25** and **26**). Both these complexes were active against the tested cell lines, including one cisplatin-resistant cell line. However, complex **25** was more active than complex **26** against each of the cell lines. The half maximal effective concentration (EC_{50}) values for complexes **25** and **26** against the cell lines murine melanoma (B16F10), epidermoid carcinoma (A431) and human ovarian carcinoma (OVCAR3) were 4.1 and 15.6 μM, 4.0 and >50 μM, and 4.2 μM and 24.5 μM, respectively. Importantly,

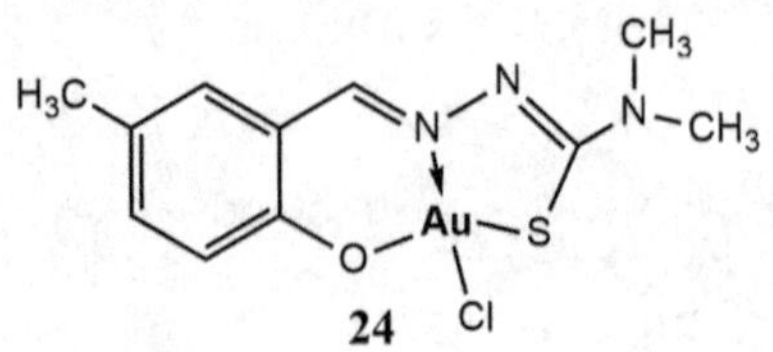

24

Figure 3.7 Chemical structures of gold(III) α-N-heterocyclic thiosemicarbazone complex **24**. Adapted from ref. 35 with permission from American Chemical Society, Copyright 2021.

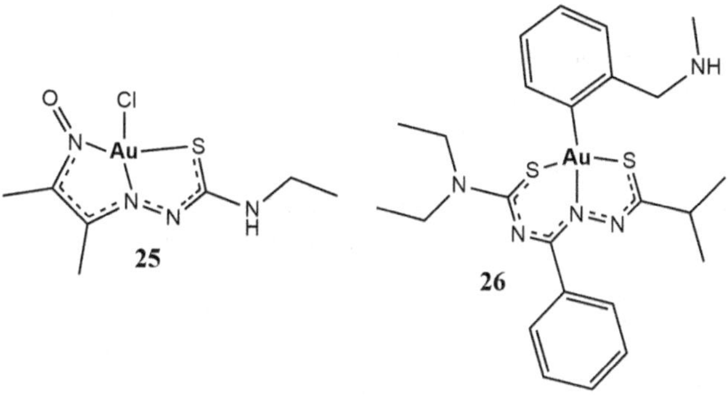

Figure 3.8 Chemical structures of Au(III)-thiosemicarbazonate complexes (**25** and **26**). Adapted from ref. 36 with permission from Elsevier, Copyright 2021.

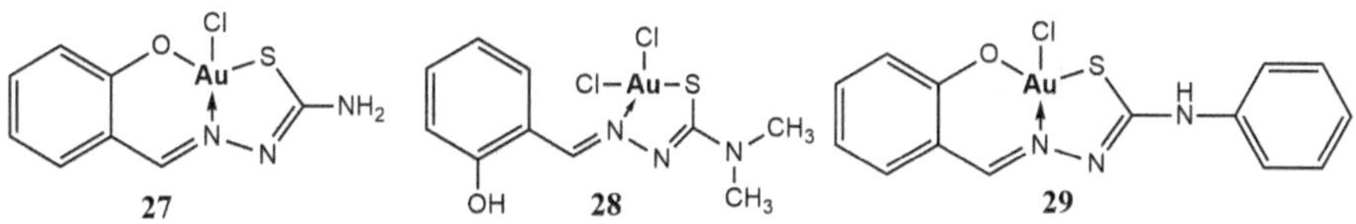

Figure 3.9 Chemical structures of Au(III)-thiosemicarbazonate complexes (**27–29**). Adapted from ref. 38 with permission from Elsevier, Copyright 2024.

complex **26** was less toxic to mouse embryonic fibroblasts (3T3) cell line (EC_{50} = 30.58 μM), compared to **25** (EC_{50} = 7.17 μM). Both the complexes displayed strong affinity towards human serum albumin. However, none of the complexes were strong DNA intercalators. A critical analysis of this research report indicates that both these complexes are exciting leads for the design and development of future anticancer drugs against less sensitive or cisplatin-resistant cancer cells.

Malignant pleural mesothelioma is a less common cancer, wherein pleural mesothelial cells are affected. Exposure to asbestos fibers is recognized as an important causative agent for this type of cancer.[37] People affected with malignant pleural mesothelioma have a low life expectancy owing to the aggressive nature of this cancer. Also, the affected pleural mesothelioma cells are therapy resistant, which further complicates the treatment of this cancer. In view of these facts, Scaccaglia *et al.*[38] reported the anticancer effects of three gold(III) complexes of thiosemicarbazone ligands (Figure 3.9, complexes **27–29**) against lung cancer cell lines, namely, A549, malignant pleural mesothelioma (H2052) and lung mesothelioma (H28). A concentration-dependent cytotoxic response was observed for the complexes, with H28 cells displaying the highest sensitivity to treatment with the complexes. The complexes induced arrest of the cell cycle and promoted apoptosis. In addition, distinct morphological changes and increased oxidative stress promoted induction of the death of the cancer cells. Importantly, the complexes suppressed interleukin-6 production in mesothelioma cells, which indicated their anti-inflammatory potential. Interestingly, the complexes did not cause hemolysis of healthy human blood cells. In addition, no detectable toxicity of the complexes was observed through toxicity assessment using the *in vivo Galleria mellonella* model. The overall findings indicate that these complexes are potential agents for the treatment of lung cancer.

In summary, many gold complexes with thiosemicarbazone ligands have displayed significant anticancer activity towards different cancer cell lines, along with appreciable selectivity towards normal cells. In addition, several complexes have exhibited anticancer effects towards otherwise resistant cancer cells. Also, some of the complexes have shown different mechanisms of action with respect to the commonly used anticancer drugs. In view of the findings described here, gold-thiosemicarbazone complexes are potential candidates for anticancer drug development.

3.5 Cyclometalated Gold Complexes

Gold(III) complexes containing C^N-cyclometalated ligands are a unique class of compounds with various applications because of their exceptional physico-chemical properties.[39] These complexes have been extensively studied for their applications in bioinorganic medicinal chemistry.[2,5,7,40] C^N-cyclometalated ligands offer greater resilience of the gold(III) ion against reduction in aqueous media, in comparison with N^N ligands, which has triggered the search for possible alternatives to platinum(II)-based anticancer agents with gold complexes. The first reports of cyclometalated Au(III) complexes with anticancer properties were based on the C^N ligand, 2-(*N*,*N*-dimethylamino)methylphenyl. These cyclometalated gold(III) complexes were found to be more cytotoxic than cisplatin against human cancer cells *in vitro*, and exhibited anticancer activity similar to cisplatin in *in vivo* studies.[41,42] These investigations were followed by reports of other cyclometalated Au(III) complexes bearing C^N chelates, in addition to C^N^N, N^C^N and C^N^C pincer ligands, with exciting anticancer activities and unique mechanisms of action.[2,5,7,40,43] One of the important advantages of cyclometalated complexes is that they tolerate a large diversity of ancillary ligands, including NHCs and phosphanes, which allows tuning of their bioactivities.[43–46] These complexes display anticancer effects through interactions with important protein targets;[2,47,48] however, interactions of these complexes with DNA are less relevant.[49]

In light of the above-mentioned features of cyclometalated gold complexes, Carboni *et al.*[50] documented the anticancer activities of a series of C^N^N-cyclometalated Au(III) complexes against A549, MCF-7, human ovarian cancer (SKOV-3) and human melanoma (A375) cancer cell lines, and normal cell line of embryonic kidney cells (HEK-293). Complex **30** (Figure 3.10) was the most active complex, with EC_{50} values in the range 1.46 ± 1.07 to 4.5 ± 0.8. Interestingly, the complex was several-fold more active than cisplatin towards the tested cell lines. However, the complex was also toxic to the non-cancerous

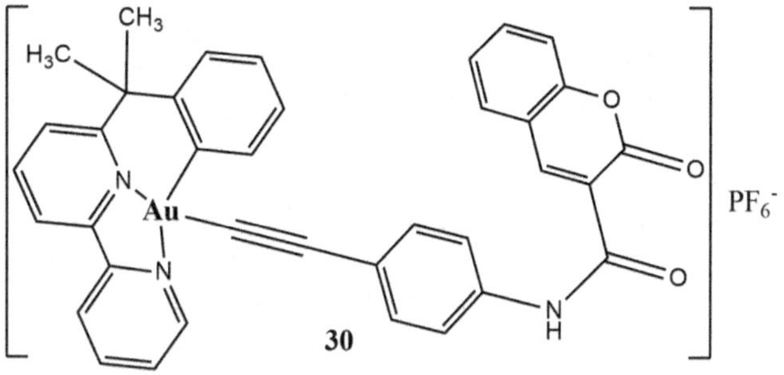

Figure 3.10 Chemical structure of C^N^N-cyclometalated Au(III) complex **30**. Adapted from ref. 50 with permission from American Chemical Society, Copyright 2018.

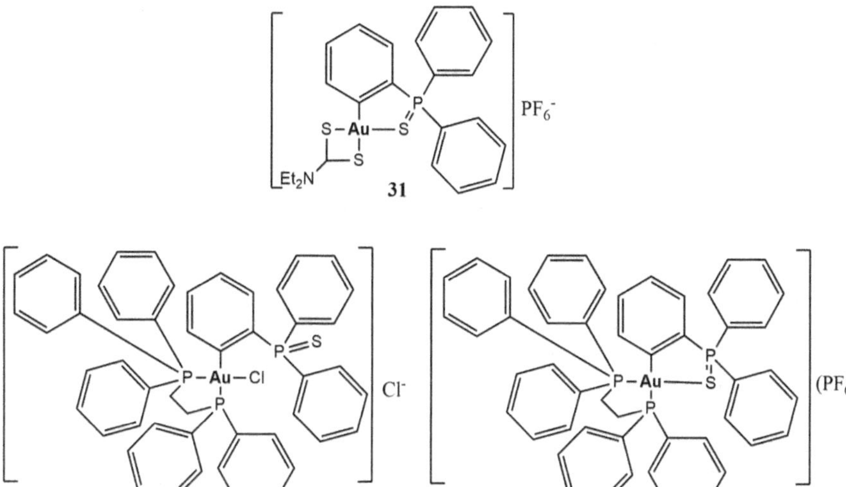

32 **33**

Figure 3.11 Chemical structures of gold(III) complexes bearing cyclometalated triphenylphosphine sulfide ligands (**31–33**). Adapted from ref. 51 with permission from American Chemical Society, Copyright 2020.

HEK-293 cells. In another study, Reddy and co-workers[51] reported *in vitro* anticancer potencies of a series of gold(III) complexes bearing cyclometalated triphenylphosphine sulfide ligands against five human cancer cell lines: cervical carcinoma (HeLa), A549, prostate cancer (PC3), fibrosarcoma (HT1080) and MDA-MB-231. Complexes **31–33** (Figure 3.11) showed strong cell growth inhibition, with IC_{50} values in the range 0.17–2.50 μM. The anticancer activities of the complexes were comparable to, or even better than, cisplatin (IC_{50} = 0.63–6.35 μM). Mechanistic investigations with HeLa cells revealed that the complexes displayed cytotoxic effects *via* apoptosis through ROS accumulation. Interestingly, substantial inhibition of endothelial cell migration and tube formation in the angiogenesis process was exhibited by complex **31**. Importantly, substantial tumor growth inhibition of 55% with 1 mg kg^{-1} dose every 3 days was observed with complex **31** in comparison with the same dose of cisplatin (tumor growth inhibition = 28%) in nude mice bearing HeLa xenografts. Overall, this research report indicated the potential of gold(III) complexes bearing cyclometalated triphenylphosphine sulfide ligands.

High tumor specificity has been reported with photo-activatable anticancer metal complexes. The activity of such complexes is triggered by the dissociation of ligands or through photochemical reduction.[52–54] Metal complexes with hydride anions as σ-donors have exhibited strong photoreactivity.[55–57] In response to the utility of photo-activatable complexes for cancer treatment, Luo and co-workers[58] reported anticancer activities of gold(III) hydride complexes (Figure 3.12, complexes **34–37**) in the dark and under visible irradiation, against human colon carcinoma (HCT116), A549, MCF-7 and hepatocellular carcinoma (HepG2) cancer cell lines. The complexes were inactive under dark conditions, with IC_{50} values greater than 100 μM after 48 h of incubation. However, after 365 nm light irradiation for 5 min, the activity increased remarkably, with IC_{50} values ranging from 0.24 to 5.86 μM, which indicated a more than 400-fold stronger activity compared to the dark conditions. By irradiating with visible at 420 nm light for 10 min, the complexes showed high activity, with IC_{50} values in the range 0.95–16.8 μM. On irradiation with longer wavelength light, at 460 nm, significant activity was observed for complexes **35, 36** and **37**. In addition, the complexes offered the

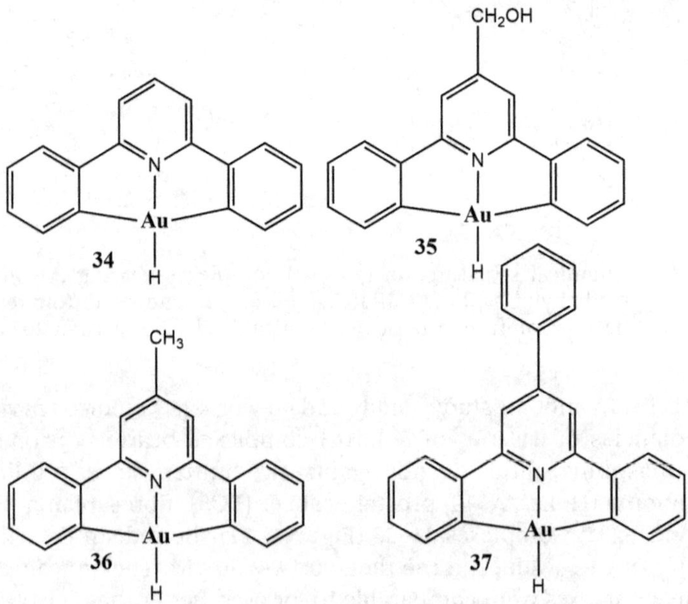

Figure 3.12 Chemical structures of gold(III)-hydride complexes (**34–37**). Adapted from ref. 58 with permission from John Wiley and Sons, Copyright 2020.

possibility to avoid the issue of drug deactivation that is generally caused by serum albumin. In addition, the complexes provided an avenue for achieving high cancer cell specificity through spatial and temporal regulation of the light irradiation. Overall, this study paves the way for the design and development of gold complexes with high anticancer activity and low side effects.

Gold complexes containing phosphine ligands with different structures have been reported with interesting anticancer properties, and have been suggested as important agents for probe development.[59,60] Berners-Price and co-workers have documented the anticancer activities of gold-phosphine complexes. The authors have also attempted to improve the *in vitro* and *in vivo* efficacy of gold-phosphine complexes.[61–64] Kim *et al.*[65] reported cyclometalated Au(III) complexes of phosphine ligands (Figure 3.13). The complexes displayed high anticancer activity against erythroleukemic (K562), non-small cell lung cancer (H460) and ovarian cancer (OVCAR8) human cancer cell lines, with IC_{50} values in the range 0.10–2.53 μM. Interestingly, the complexes were several-fold more active than cisplatin. The complexes triggered early late-stage apoptosis *via* disruption of redox homeostasis. Complexes **38** and **40** induced predominantly G1 cell cycle arrest.

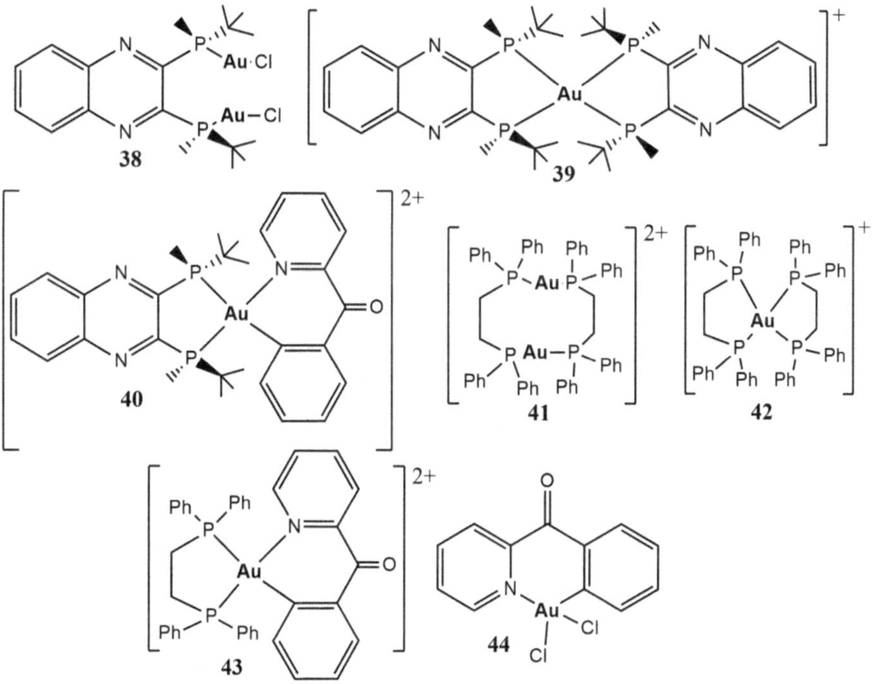

Figure 3.13 Chemical structures of gold(III)-phosphine complexes (**38–44**). Adapted from ref. 65, https://doi.org/10.1038/s41598-019-48584-5, under the terms of the CC BY 4.0 license, http://creativecommons.org/licenses/by/4.0/.

Bisphosphines represent a diverse class of bidentate ligands with extensive uses in important biological systems.[66] Earlier research on anticancer gold complexes containing bisphosphine ligands[61,63,67–69] has revealed their significance as possible anticancer agents. The effect of the nature of phosphine ligands on the anticancer activities of phosphinogold(I) complexes was independently investigated by Darkwa *et al.* and Raubenheimer and co-workers. Both these research groups documented that the anticancer activity of the complexes was dependent on the substituents on the phosphorus atom of the phosphine ligand. It was observed that bisphosphine complexes containing longer CH_2 linkers had better anticancer activities than complexes containing shorter CH_2 linkers.[70,71] In view of these observations, Arojoye *et al.*[72] reported the anticancer efficacy of cationic biphenyl organogold(III) complexes supported by bisphosphine ligands against human breast cancer (MDA-MB-468), MDA-MB-231, mice breast cancer (4T1) and glioblastoma (BT-333) cancer cells. Complexes **45–48** (Figure 3.14) showed promising anticancer activities against the tested cell lines. It is important to note that the complexes were more active than cisplatin against the tested cell lines. In addition, complex **47** showed remarkable tumor growth inhibition in a metastatic triple-negative breast cancer (TNBC) mouse model *in vivo*. In addition,

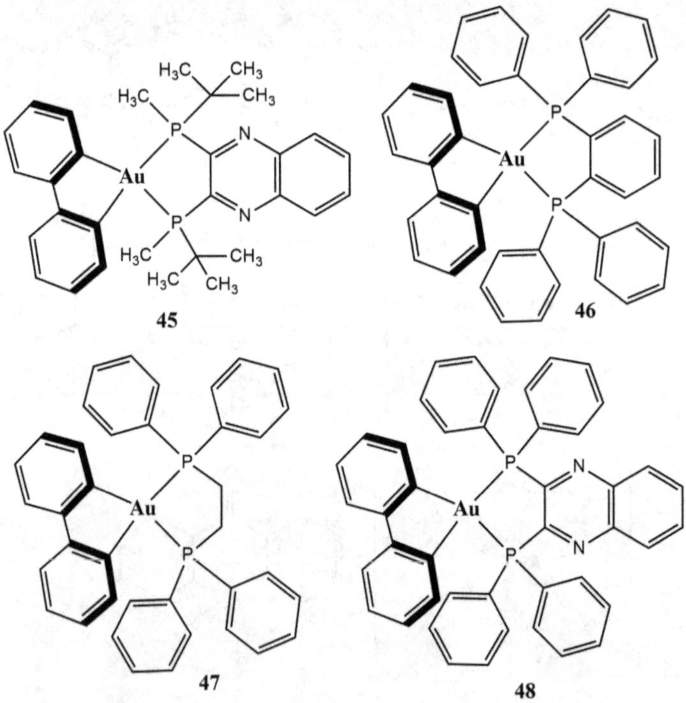

Figure 3.14 Chemical structures of cationic biphenyl-organogold(III) complexes containing bisphosphine ligands (**45–48**). Adapted from ref. 72 with permission from American Chemical Society, Copyright 2023.

complex **47** exhibited appreciable blood serum stability over a period of 24 h. The anticancer effects of complex **47** were due to the induction of mitochondrial uncoupling, membrane depolarization and G1 cell cycle arrest. Overall, this research report showed the tremendous potential of Au(III) as a lead for the design and development of gold-phosphine-based anticancer complexes.

Analysis of the research reports discussed in this section indicate that several cyclometalated gold complexes have demonstrated exciting anticancer activities (greater than that for cisplatin) towards different cancer cell lines, along with selectivity towards normal cells. In addition, several complexes are potentially photo-activatable. Additionally, several complexes have displayed anticancer activities towards resistant cancer cells. Also, some of complexes have displayed anticancer effects through mechanisms of action that are different to those seen in commonly used anticancer drugs. In view of the findings described here, cyclometalated gold complexes are potential candidates for anticancer drug design and development.

3.6 Multinuclear Gold Complexes

Generally, multinuclear complexes display novel DNA-binding modes, which leads to long-range DNA cross-links, bis-intercalation, phosphate clamps, interduplex cross-links and DNA–protein cross-links.[73] Multinuclear complexes seem promising for the design and development of anticancer drugs, as the presence of two or more active metals is thought to enhance the interactions of such complexes with therapeutically important target biomolecules, such as nucleic acids, proteins, enzymes, *etc.* Recently, multinuclear gold complexes have been investigated as anticancer agents with strong antiproliferative properties.[74,75] Lescure *et al.*[76] reported the anticancer properties of three near-infrared homobimetallic gold(I)-based aza-BODIPY complexes, **49–51** (Figure 3.15) against human and murine cancer cell lines, namely, 4T1, MDA-MB-231, colon carcinoma (CT26) and colon adenocarcinoma (SW480). The complexes showed promising *in vitro* anticancer activity against the tested cell lines in low micromolar range. However, none of the complexes showed better activity than auranofin. Interestingly, the complexes were trackable through near-infrared imaging in cells and in mice.

Improved pharmacokinetic and pharmodynamic profiles of metal complexes, along with enhanced selectivity for different biological targets, have been reported in heterometallic multinuclear complexes. The improvements have been attributed to the synergism between the metal centers within the complexes. This approach offers a possible mechanism for overcoming the resistance to anticancer drugs by cancer cells.[77] Niekerk *et al.*[78] reported anticancer activities and toxicities of two dyads containing a gold-porphyrin affixed to organotin(IV) entities (Figure 3.16, complexes **52** and **53**) towards MCF-7 cells and healthy human fibroblasts (FS 20–68), respectively. The dyads were observed to be more active than the mixture of the individual components. Synergism of the anticancer effects of the

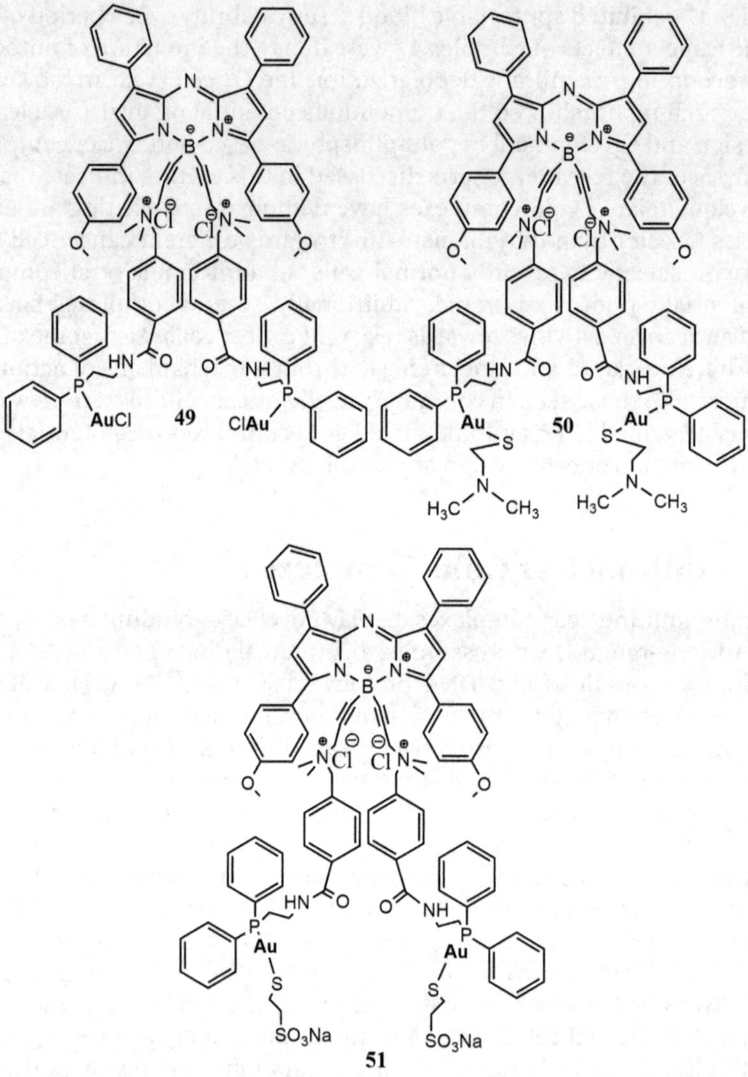

Figure 3.15 Chemical structures of near-infrared homobimetallic gold(I)-based aza-BODIPY complexes (**49–51**). Adapted from ref. 76 with permission from Elsevier, Copyright 2021.

two components was observed due to the covalent linkage of the organotin entity with the gold-porphyrin system. The dyad showed high anticancer activity (IC_{50} = 0.13 μM) against MCF-7 cells, along with appreciable selectivity towards healthy human fibroblasts. Overall, this research revealed the benefits of the combination of gold-porphyrins with organotin moieties for anticancer drug development. Recently, Quintana and co-workers[79] reported the anticancer activities of two dinuclear gold(I) complexes (Figure 3.17, complexes **54** and **55**) towards the human gastric cancer

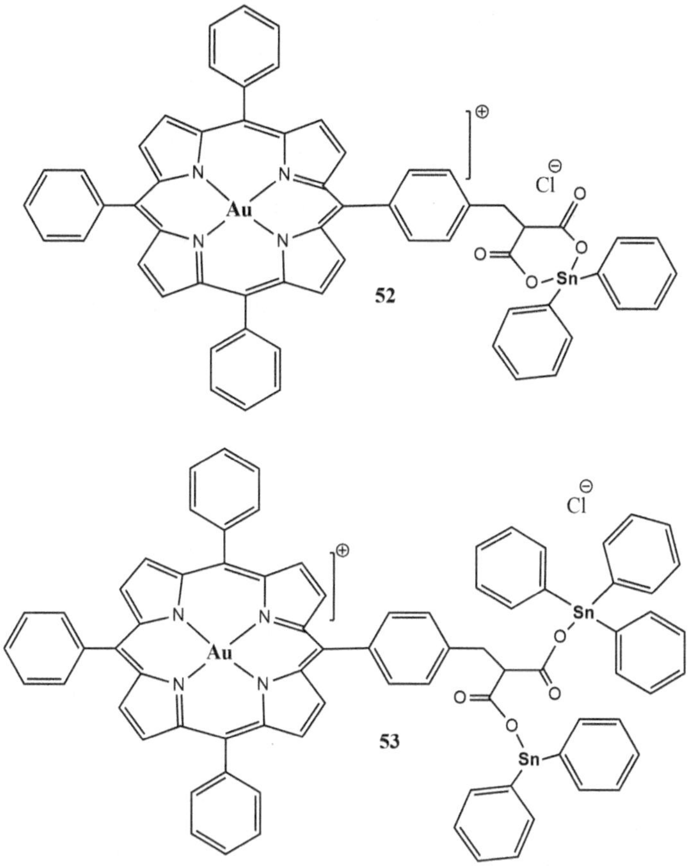

52

53

Figure 3.16 Chemical structures of dyads containing a gold-porphyrin affixed to organotin(IV) entities (**52** and **53**). Adapted from ref. 77 with permission from John Wiley and Sons, Copyright 2019.

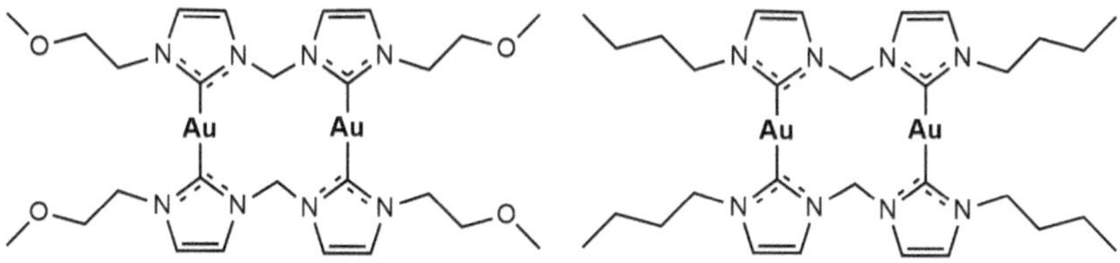

54 55

Figure 3.17 Chemical structures of dinuclear gold(I) complexes (**54** and **55**). Adapted from ref. 79 with permission from Elsevier, Copyright 2022.

cell line (HGC-27), and the non-tumorigenic Madin–Darby canine kidney epithelial (MDCK) cell line. Complex **55** displayed substantial anticancer activity against the HGC-27 cell line. Complex **55** was further investigated against five human ovarian carcinoma cell lines, namely, A2780, IGROV1, OVCAR3, OVCAR4 and SKOV3, and six human colon carcinoma cell lines, namely, CACO2, COLO205, HCT116, HCT15, SW620 and HT-29, in addition to the non-cancerous African green monkey kidney cells (Vero) and resting, non-proliferative peripheral blood mononuclear cells. The complex exhibited strong anticancer activity against the A2780 cell line. Complex **55** was found to enhance oxidative stress by triggering ROS production and preventing the scavenging activity of TrxR.

The research discussed in this section indicates that multinuclear gold complexes have demonstrated promising anticancer activities towards different cancer cell lines, in addition to being selective towards normal cells. In addition, several complexes are photo-activatable and can be activated at the target site through proper irradiation. Moreover, certain heterometallic complexes have displayed synergism of their activities between the different metal centers. From the findings discussed here, multinuclear gold complexes are potential anticancer drug candidates and thus warrant further investigations.

3.7 Complexes with Bioactive Ligands

For the development of more active anticancer agents, a hybrid medicinal chemistry approach has been employed by several researchers. In this approach, metal complexes are either linked with biologically active organic frameworks or ligands, or the complexes simply contain biologically active ligands. The literature describes several examples of metal–organic scaffolds containing well-known drug molecules with promising anticancer properties.[80–84] Bian *et al.*[85] reported a gold(I) complex bearing an oleanolic acid derivative (Figure 3.18, compound **56**) that induced apoptosis in A2780 cells by activation of endoplasmic reticulum stress (ERS). The complex inhibited tumor growth of the A2780 xenograft *via* increasing ERS levels and decreasing TrxR activity in tumor tissues.

Recently, artemisinin and several of its derivatives have been explored as treatment options for different diseases, including cancer.[86,87] Activation

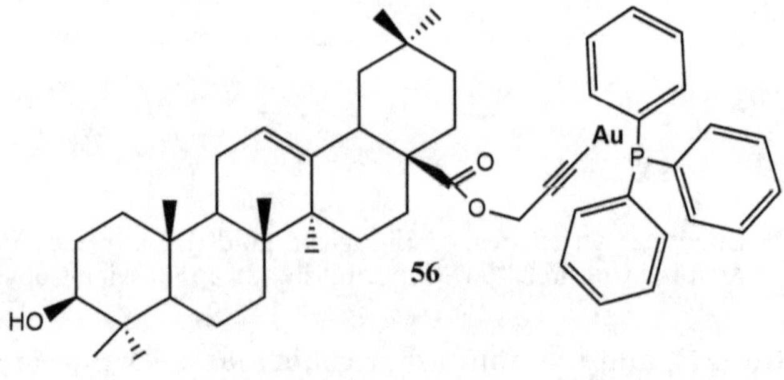

56

Figure 3.18 Chemical structure of gold(I) complex bearing an oleanolic acid derivative (**56**). Adapted from ref. 85 with permission from John Wiley and Sons, Copyright 2020.

of artemisinin derivatives by iron from free heme leads to ROS generation, which is thought to be the possible mechanism for the anticancer effects of artemisinin and its derivatives.[88–90] Mitochondria are the preferred sites for the activation of artemisinin derivatives, as fresh heme is continuously biosynthesized there. Furthermore, it has been reported that artemisinin derivatives that target mitochondria have better anticancer efficacy, in comparison to non-mitochondria-targeting artemisinin derivatives.[91] Recently, Zhang *et al.*[92] reported hybrid complexes combining the cationic bis(NHC)-gold(I) unit with an artemisinin derivative. The hybrid complex **57** (Figure 3.19) displayed robust anticancer activities towards several human cancer cell lines, including breast, lung, prostate, bladder, bone, liver and acute and chronic myeloid leukemias, with 50% growth inhibition (GI_{50}) values in the nanomolar range. In addition, complex **57** showed a high selectivity towards the cancer cells in comparison to normal cells. The complex inhibited the redox antioxidant transcription factor nuclear factor erythroid 2-related factor 2 (NRF2), which usually leads to aggressiveness of cancer cells and their resistance to anticancer drugs. Complex **57** sensitized sorafenib in HepG2 cells, wherein dysregulated NRF2 signaling is related to primary and acquired drug resistance. In addition, nuclear factor kappa B (NF-κB) and hypoxia-inducible factor transcriptional activities, which are also linked with progression and resistance in cancer, were also inhibited by complex **57**. This research conclusively reported hybrid gold(I) complexes as a new class of metal complexes that could lead to the development of new and efficient anticancer drugs.

Papaverine is an alkaloid used for treating impotence and vasospasms.[93–95] In addition, papaverine is known to display antiviral properties against several viruses, such as cytomegalovirus, measles, respiratory syncytial virus and HIV.[96] Papaverine has also exhibited selective anticancer properties against

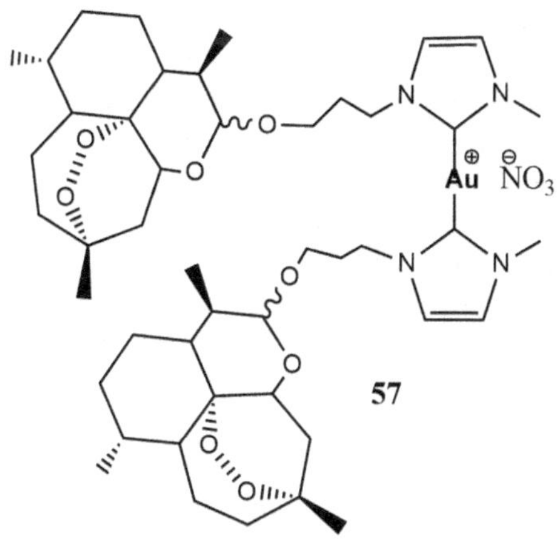

57

Figure 3.19 Chemical structure of cationic bis(NHC)-gold(I) complex (**57**). Adapted from ref. 92 with permission from John Wiley and Sons, Copyright 2020.

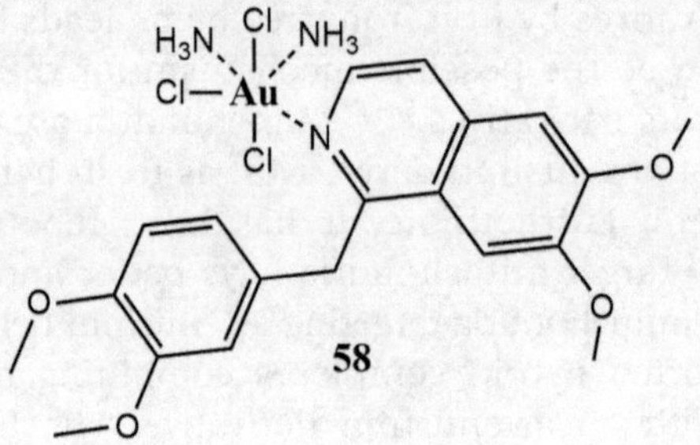

58

Figure 3.20 Chemical structure of papaverine-gold(III) complex (**58**). Adapted from ref. 102, https://doi.org/10.3390/molecules25225447, under the terms of the CC BY 4.0 license, http://creativecommons.org/licenses/by/4.0/.

several cancer cell lines, such as MCF-7 and MDA-MB-231,[97] breast carcinoma (T47D),[96] colorectal carcinoma (HT29),[96] HT1080,[96] PC-3,[98] human prostate carcinoma (LNCaP)[99,100] and HepG2.[101] In response to the interesting therapeutic efficacy of papaverine, Gaber and co-workers[102] reported enhancement of the anticancer properties of a papaverine-gold(III) complex (Figure 3.20, complex **58**) against MCF-7 and HepG2 cells. The authors reported that the anticancer properties of the complex were better than those of papaverine. In addition, the anticancer activity of the complex against MCF-7 cells was higher than for the reference drug cisplatin. It is important to mention here that the complex was less toxic towards normal cells than papaverine, with an IC_{50} value of 111 μg mL^{-1}. Owing to the results obtained, the papaverine-Au(III) complex can be regarded as a promising anticancer gold complex, which should be further evaluated *in vivo*.

Metformin has been used for the treatment of type 2 diabetes for the last 50 years. At present, metformin is being investigated as an anticancer agent owing to its chemopreventive properties.[103–105] Metformin targets mitochondria by causing inhibition of mitochondrial complex I, which leads to energy stress and thereby induces cytotoxicity in cancer cells. Taking advantage of the therapeutic properties of metformin, Kim *et al.*[106] reported anticancer activities of stable and water-soluble gold(III) complexes of metformin or phenylmetformin (Figure 3.21, compounds **59–62**) against MDA-MB-231, MDA-MB-468 and H460 cancer cell lines. The complexes showed substantial mitochondrial uptake, and induced anticancer activities in the micromolar range in MDA-MB-231 and MDA-MB-468 cells. The anticancer activities of the complexes were many times higher than for metformin or phenylmetformin alone.

Tamoxifen has been used for treating estrogen receptor-positive breast cancer at all stages of the disease. It has been listed as an essential drug by the World Health Organization for the endocrine treatment of breast cancer. It is the first cancer chemopreventive drug approved by the US FDA for reducing the incidence of breast cancer in pre- and post-menopausal women at high risk.[107] To synergize the anticancer potency of tamoxifen with gold

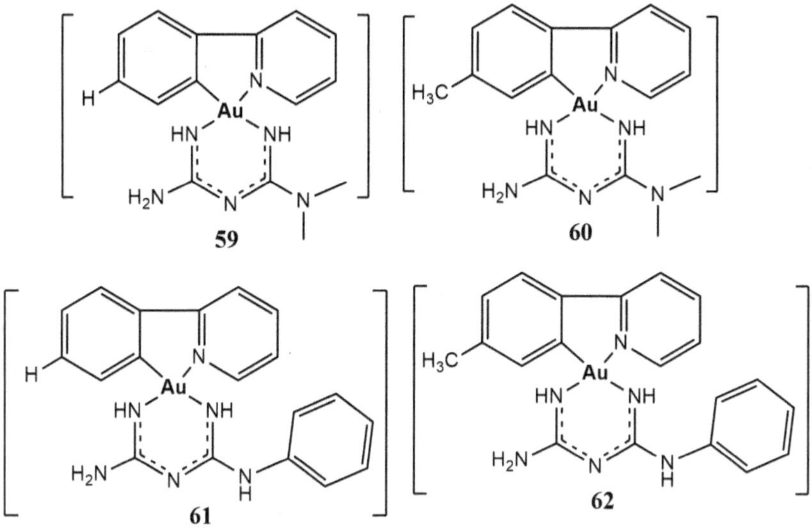

59 60 61 62

Figure 3.21 Chemical structure of water-soluble gold(III) complexes of metformin and phenylmetformin (**59–62**). Adapted from ref. 106 with permission from John Wiley and Sons, Copyright 2021.

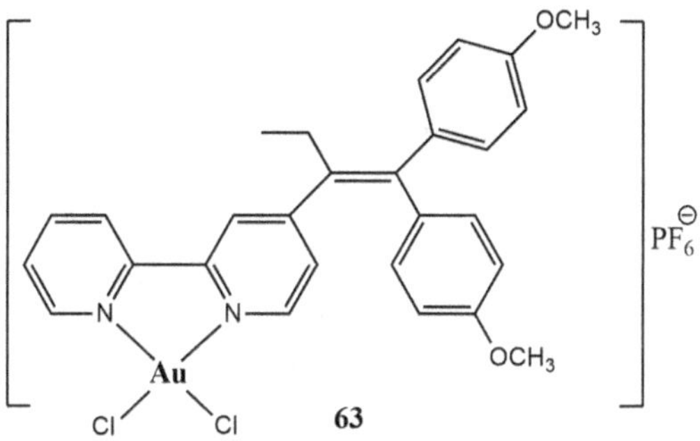

63

Figure 3.22 Chemical structure of Au(III) complex of tamoxifen-derived pharmacophore (**63**). Adapted from ref. 108, https://doi.org/10.1021/acs.jmedchem.3c00617, under the terms of the CC BY 4.0 license, http://creativecommons.org/licenses/by/4.0/.

complexes, Scalcon *et al.*[108] recently prepared a Au(III) complex of a tamoxifen-derived pharmacophore (Figure 3.22, complex **63**). The complex was active against MCF-7 and MDA-MB 231 cancer cells. Importantly, the complex was more active against ER^+ (MCF-7) cells *vs.* the ER^- (MDA-MB 231) cells. In addition, the complex showed higher activity than cisplatin under the tested conditions. The binding ability of the complex to the estrogen receptor was retained and the complex inhibited the seleno enzyme TrxR. Redox imbalance, along with a reduction in total thiols and augmented ROS production was noticed in breast cancer cells upon treatment with the complex.

Interestingly, the complex induced mitochondrial damage by affecting mitochondrial respiration, membrane potential and mitochondrial morphology.

A critical assessment of the research described in this section indicates that several types of therapeutically important ligands and ligand systems have been used for designing and developing anticancer gold complexes. In almost all of the reports, enhancement of the anticancer properties of the gold complexes with bioactive ligands has been reported. In several reports the hybrid complexes have shown much better therapeutic profiles than the bioactive ligand and the reference drug. In addition, the complexes have displayed strong and interesting mechanisms of action against the cancer cell lines tested. Overall, gold complexes of bioactive ligands appear as potential candidates for the development of anticancer gold complexes, and thus further drug design rationales and investigations are warranted.

References

1. S. Nobili, E. Mini, I. Landini, C. Gabbiani, A. Casini and L. Messori, *Med. Res. Rev.*, 2010, **30**, 550.
2. B. Bertrand and A. Casini, *Dalton Trans.*, 2014, **43**, 4209.
3. C. Nardon, G. Boscutti and D. Fregona, *Anticancer Res.*, 2014, **34**, 487.
4. P. I. Da Silva Maia, V. M. Deflon and U. Abram, *Future Med. Chem.*, 2014, **6**, 1515.
5. T. Zou, C. T. Lum, C.-N. N. Lok, J.-J. J. Zhang and C.-M. M. Che, *Chem. Soc. Rev.*, 2015, **44**, 8786.
6. T. Lazarević, A. Rilak and Ž. D. Bugarčić, *Eur. J. Med. Chem.*, 2017, **142**, 8.
7. B. Bertrand, M. R. M. Williams and M. Bochmann, *Chem. – Eur. J.*, 2018, **24**, 11840.
8. C. I. Yeo, K. K. Ooi and E. R. T. Tiekink, *Molecules*, 2018, **23**, 1410.
9. R. P. Herrera and M. C. Gimeno, *Chem. Rev.*, 2021, **121**, 8311.
10. C. Nardon, N. Pettenuzzo and D. Fregona, *Curr. Med. Chem.*, 2016, **23**, 3374.
11. E. E. Langdon-Jones and S. J. A. A. Pope, *Chem. Commun.*, 2014, **50**, 10343.
12. G. Faa, C. Gerosa, D. Fanni, J. I. Lachowicz and V. M. Nurchi, *Curr. Med. Chem.*, 2018, **25**, 75.
13. J. H. Kim, E. Reeder and S. Parkin, *Sci. Rep.*, 2019, **9**, 12335.
14. L. Ronconi, C. Marzano, P. Zanello, M. Corsini, G. Miolo, C. Maccà, A. Trevisan and D. Fregona, *J. Med. Chem.*, 2006, **49**, 1648.
15. K. C. Tong, D. Hu, P. K. Wan, C. N. Lok and C. M. Che, *Front. Chem.*, 2020, **8**, 587207.
16. B. Bertrand, L. Stefan, M. Pirrotta, D. Monchaud, E. Bodio, P. Richard, P. Le Gendre, E. Warmerdam, M. H. de Jager, G. M. Groothuis, M. Picquet and A. Casini, *Inorg. Chem.*, 2014, **53**, 2296.
17. P. Hikisz, Ł. Szczupak, A. Koceva-Chyła, A. Gu Spiel, L. Oehninger, I. Ott, B. Therrien, J. Solecka and K. Kowalski, *Molecules*, 2015, **20**, 19699.
18. I. Ali, W. A. Wani, K. Saleem and M.-F. Hsieh, *Polyhedron*, 2013, **56**, 134.

19. M. Altaf, A. A. Isab, J. Vančo, Z. Dvořák, Z. Trávníček and H. Stoeckli-Evans, *RSC Adv.*, 2015, **5**, 81599.
20. A. Pintus, M. C. Aragoni, M. A. Cinellu, L. Maiore, F. Isaia, V. Lippolis, G. Orrù, E. Tuveri, A. Zucca and M. Arca, *J. Inorg. Biochem.*, 2017, **170**, 188.
21. S. A. Sousa, J. H. Leitão, R. A. L. Silva, D. Belo, I. C. Santos, J. F. Guerreiro, M. Martins, D. Fontinha, M. Prudêncio, M. Almeida, D. Lorcy and F. Marques, *J. Inorg. Biochem.*, 2020, **202**, 10904.
22. D. Fontinha, S. A. Sousa, T. S. Morais, M. Prudencio, J. H. Leitao, Y. Le Gal, D. Lorcy, R. A. L. Silva, M. F. G. Velho, D. Belo, M. A. Joana, F. Guerreiro, T. Pinheiro and F. Marques, *Metallomics*, 2020, **12**, 974.
23. S. A. Patil, S. A. Patil, R. Patil, R. S. Keri, S. Budagumpi, G. R. Balakrishna and M. Tacke, *Future Med. Chem.*, 2015, **7**, 1305.
24. F. Cisnetti and A. Gautier, *Angew. Chem., Int. Ed.*, 2013, **52**, 11976.
25. W. A. Herrmann, *Angew. Chem., Int. Ed.*, 2002, **41**, 1290–1309.
26. F. E. Hahn and M. C. Jahnke, *Angew. Chem., Int. Ed.*, 2008, **47**, 3122–3172.
27. E. Peris and R. H. Crabtree, *Coord. Chem. Rev.*, 2004, **248**, 2239–2246.
28. J. C. Garrison and W. J. Youngs, *Chem. Rev.*, 2005, **105**, 3978–4008.
29. A. T. Termaten, M. Schakel, A. W. Ehelers, M. Lutz, A. L. Spek and K. Lammertsma, *Chem. – Eur. J.*, 2003, **9**, 3577–3582.
30. W. A. Herrmann, L. J. Goossen, C. Kocher and G. R. J. Artus, *Angew. Chem., Int. Ed.*, 1996, **35**, 2805–2807.
31. U. Kızrak, O. C. I. Ozdemir, N. Gürbüz, S. D. Düsünceli, M. Kaloglu, L. Mansour, F. Zaghrouba, N. Hamdi and I. Ozdemir, *J. Organomet. Chem.*, 2019, **882**, 26.
32. S. Sen, Y. Li, V. Lynch, K. Arumugam, J. L. Sessler and J. F. Arambula, *Chem. Commun.*, 2019, **55**, 10627.
33. S. R. Jean, D. V. Tulumello, C. Riganti, S. U. Liyanage, A. D. Schimmer and S. O. Kelley, *ACS Chem. Biol.*, 2015, **10**, 2007.
34. R. Büssing, B. Karge, P. Lippmann, P. G. Jones, M. Brönstrup and I. Ott, *ChemMedChem*, 2021, **16**, 1.
35. J. Zhang, M. Jiang, S. Li, Z. Zhang, H. Sun, F. Yang and H. Liang, *J. Med. Chem.*, 2021, **64**, 6777.
36. B. Possato, L. F. Dalmolin, L. M. Pereira, J. Q. Alves, R. T. C. Silva, R. V. Gelamo, A. P. Yatsuda, R. F. V. Lopez, S. de Albuquerque, N. B. Leite and P. I. D. S. Maia, *Eur. J. Pharm. Sci.*, 2021, **162**, 105834.
37. E. Schillebeeckx, J. P. van Meerbeeck and K. Lamote, *Eur. Respir. Rev.*, 2023, **30**, 210057.
38. M. Scaccaglia, S. Pinelli, L. Manini, B. Ghezzi, M. Nicastro, J. Heinrich, N. Kulak, P. Mozzoni, G. Pelosi and F. Bisceglie, *J. Inorg. Biochem.*, 2024, **251**, 112438.
39. R. Kumar and C. Nevado, *Angew. Chem., Int. Ed.*, 2017, **56**, 1994.
40. S. Jurgens, F. E. Kuhn and A. Casini, *Curr. Med. Chem.*, 2018, **25**, 437.
41. R. G. Buckley, A. M. Elsome, S. P. Fricker, G. R. Henderson, B. R. C. Theobald, R. V. Parish, B. P. Howe and L. R. Kelland, *J. Med. Chem.*, 1996, **39**, 5208.

42. R. V. Parish, B. P. Howe, J. P. Wright, J. Mack, R. G. Pritchard, R. G. Buckley, A. M. Elsome and S. P. Fricker, *Inorg. Chem.*, 1996, **35**, 1659.
43. S. Jurgens, V. Scalcon, N. Estrada-Ortiz, A. Folda, F. Tonolo, C. Jandl, D. L. Browne, M. P. Rigobello, F. E. Kuhn and A. Casini, *Bioorg. Med. Chem.*, 2017, **25**, 54520.
44. J. J. Yan, A. L.-F. Chow, C.-H. Leung, R. W.-Y. Sun, D.-L. Ma and C.-M. Che, *Chem. Commun.*, 2010, **46**, 3893.
45. B. Bertrand, S. Spreckelmeyer, E. Bodio, F. Cocco, M. Picquet, P. Richard, P. Le Gendre, C. Orvig, M. Cinellu and A. Casini, *Dalton Trans.*, 2015, **44**, 11911.
46. R. W.-Y. Sun, C.-N. Lok, T. T.-H. Fong, C. K.-L. Li, Z. F. Yang, T. Zou, A. F.-M. Siu and C.-M. Che, *Chem. Sci.*, 2013, **4**, 1979.
47. M. N. Wenzel, S. M. Meier-Menches, T. L. Williams, E. Rämisch, G. Barone and A. Casini, *Chem. Commun.*, 2018, **54**, 611.
48. C. Gabbiani, G. Mastrobuoni, F. Sorrentino, B. Dani, M. P. Rigobello, A. Bindoli, M. A. Cinellu, G. Pieraccini, L. Messori and A. Casini, *MedChemComm*, 2011, **2**, 50.
49. A. Casini, R. Sun and I. Ott, in *Metallo-Drugs: Development and Action of Anticancer Agents*, De Gruyter, 2018, vol. 18, pp. 199–218.
50. S. Carboni, A. Zucca, S. Stoccoro, L. Maiore, M. Arca, F. Ortu, C. Artner, B. K. Keppler, S. M. Meier-Menches and A. Casini, *Inorg. Chem.*, 2018, **57**, 14852.
51. T. Srinivasa Reddy, S. H. Priver, N. Mirzadeh, R. B. Luwor, V. Ganga Reddy, S. Ramesan and S. K. Bhargava, *Inorg. Chem.*, 2020, **59**, 5662.
52. S. H. Askes, A. Bahreman and S. Bonnet, *Angew. Chem.*, 2014, **126**, 1047.
53. L. N. Lameijer, D. Ernst, S. L. Hopkins, M. S. Meijer, S. H. Askes, S. E. Le Dévédec and S. Bonnet, *Angew. Chem.*, 2017, **129**, 11707.
54. A. Li, C. Turro and J. J. Kodanko, *Chem. Commun.*, 2018, **54**, 1280.
55. R. N. Perutz and B. Procacci, *Chem. Rev.*, 2016, **116**, 8506.
56. A. J. Jordan, G. Lalic and J. P. Sadighi, *Chem. Rev.*, 2016, **116**, 8318.
57. E. S. Wiedner, M. B. Chambers, C. L. Pitman, R. M. Bullock, A. J. Miller and A. M. Appel, *Chem. Rev.*, 2016, **116**, 8655.
58. H. Luo, B. Cao, A. S. C. Chan, R. W.-Y. Sun and T. Zou, *Angew. Chem., Int. Ed.*, 2020, **59**, 11046.
59. S. Urig, K. Fritz-Wolf, R. Réau, C. Herold-Mende, K. Tóth, E. Davioud-Charvet and K. Becker, *Angew. Chem., Int. Ed.*, 2006, **45**, 1881.
60. J. D. Wilton-Ely, A. Schier, N. W. Mitzel and H. Schmidbaur, *J. Chem. Soc., Dalton Trans.*, 2001, 1058.
61. S. J. Berners-Price, C. K. Mirabelli, R. K. Johnson, M. R. Mattern, F. L. McCabe, L. F. Faucette, C.-M. Sung, S.-M. Mong, P. J. Sadler and S. T. Crooke, *Cancer Res.*, 1986, **46**, 5486.
62. S. J. Berners-Price, R. E. Norman and P. J. Sadler, *J. Inorg. Biochem.*, 1987, **31**, 197.
63. S. J. Berners-Price, G. R. Girard, D. T. Hill, B. M. Sutton, P. S. Jarrett, L. F. Faucette, R. K. Johnson, C. K. Mirabelli and P. J. Sadler, *J. Med. Chem.*, 1990, **33**, 1386.

64. S. J. Berners-Price and P. J. Sadler, *J. Inorg. Biochem.*, 1987, **31**, 267.
65. J. H. Kim, E. Reeder, S. Parkin and S. G. Awuah, *Sci. Rep.*, 2019, **9**, 12335.
66. M. S. S. Adam, A. D. Mohamad, P. G. Jones, M. K. Kindermann and J. W. Heinicke, *Polyhedron*, 2013, **50**, 101.
67. C. K. Mirabell, R. K. Johnson, D. T. Hill, L. F. Faucette, G. R. Girard, G. Y. Kuo, C. M. Sung and S. T. Crooke, *J. Med. Chem.*, 1986, **29**, 218.
68. C. K. Mirabelli, D. T. Hill, L. F. Faucette, F. L. McCabe, G. R. Girard, D. B. Bryan, B. M. Sutton, J. O. L. Barus, S. T. Crooke and R. K. Johnson, *J. Med. Chem.*, 1987, **30**, 2181.
69. S. J. Berners-Price and P. J. Sadler, *Inorg. Chem.*, 1986, **25**, 3822.
70. F. K. Keter, I. A. Guzei, M. Nell, W. E. v. Zyl and J. Darkwa, *Inorg. Chem.*, 2014, **53**, 2058.
71. U. E. Horvath, L. Dobrzańska, C. E. Strasser, W. Bouwer, G. Joone, C. E. J. van Rensburg, S. Cronje and H. G. Raubenheimer, *J. Inorg. Biochem.*, 2012, **111**, 80.
72. A. S. Arojojoye, C. Olelewe, S. Gukathasan, J. H. Kim, H. Vekaria, S. Parkin, P. G. Sullivan and S. G. Awuah, *J. Med. Chem.*, 2023, **66**, 7868.
73. K. Wang and E. Gao, *Anti-Cancer Agents Med. Chem.*, 2014, **14**, 147.
74. A. Trommenschlager, F. Chotard, B. Bertrand, S. Amor, L. Dondaine, M. Picquet, P. Richard, A. Bettaïeb, P. Le Gendre and C. Paul, *Dalton Trans.*, 2017, **46**, 8051.
75. T. Zou and T. Zou, *Anti-Cancer N-Heterocyclic Carbene Complexes of Gold (iii), Gold (I) and Platinum (II) Thiol "Switch-On" Fluorescent Probes, Thioredoxin Reductase Inhibitors and Endoplasmic Reticulum Targeting Agents*, 2016, p. 101.
76. R. Lescure, M. Privat, J. Pliquett, A. Massot, O. Baffroy, B. Busser, P.-S. Bellaye, B. Collin, F. Denat and A. Bettaïeb, *Eur. J. Med. Chem.*, 2021, **220**, 113483.
77. A. van Niekerk, P. Chellan and S. F. Mapolie, *Eur. J. Inorg. Chem.*, 2019, **30**, 3432.
78. C. Nguyen, I. Toubia, S. Diring, K. Hadj-Kaddour, M. Gary-Bobo, M. Kobeissi and F. Odobel, *Dalton Trans.*, 2021, **50**, 4583.
79. M. Quintana, A. Rodriguez-Rius, A. Vellé, S. Vives, P. J. S. Miguel and G. Triola, *Bioorg. Med. Chem.*, 2022, **67**, 116814.
80. A. Trommenschlager, F. Chotard, B. Bertrand, S. Amor, P. Richard, A. Bettaïeb, C. Paul, J. L. Connat, P. Le Gendre and E. Bodio, *ChemMedChem*, 2018, **13**, 2408.
81. R. G. Kenny and C. J. Marmion, *RSC Metallobio. Ser.*, 2019, **14**, 3.
82. E. Meggers, *Angew. Chem., Int. Ed.*, 2011, **50**, 2442.
83. W. H. Ang, L. J. Parker, A. De Luca, L. Juillerat-Jeanneret, C. J. Morton, M. Lo Bello, M. W. Parker and P. J. Dyson, *Angew. Chem., Int. Ed.*, 2009, **48**, 3854.
84. D. Havrylyuk, A. C. Hachey, A. Fenton, D. K. Heidary and E. C. Glazer, *Nat. Commun.*, 2022, **13**, 3636.
85. M. Bian, Y. Sun, Y. Liu, Z. Xu, R. Fan, Z. Liu and W. Liu, *Chem. – Eur. J.*, 2020, **26**, 7092.

86. S. Krishna, L. Bustamante, R. K. Haynes and H. M. Staines, *Trends Pharmacol. Sci.*, 2008, **29**, 520.
87. T. Efferth, *Biotechnol. Adv.*, 2018, **36**, 1730.
88. Y. K. Wong, C. Xu, K. A. Kalesh, Y. He, Q. Lin, W. F. Wong, H. M. Shen and J. Wang, *Med. Res. Rev.*, 2017, **37**, 1492.
89. N. Eling, L. Reuter, J. Hazin, A. Hamacher-Brady and N. R. Brady, *Oncoscience*, 2015, **2**, 517.
90. E. Ooko, M. E. Saeed, O. Kadioglu, S. Sarvi, M. Colak, K. Elmasaoudi, R. Janah, H. J. Greten and T. Efferth, *Phytomedicine*, 2015, **22**, 1045.
91. C. J. Zhang, J. Wang, J. Zhang, Y. M. Lee, G. Feng, T. K. Lim, H. M. Shen, Q. Lin and B. Liu, *Angew. Chem., Int. Ed.*, 2016, **55**, 13770.
92. C. Zhang, P.-Y. Fortin, G. Barnoin, X. Qin, X. Wang, A. F. Alvarez, C. Bijani, M.-L. Maddelein, C. Hemmert, O. Cuvillier and H. Gornitzka, *Angew. Chem., Int. Ed.*, 2020, **59**, 12062.
93. J. K. Liu and W. T. Couldwell, *Neurocrit. Care*, 2005, **2**, 124.
94. C. M. Sayama, J. K. Liu and W. T. Couldwell, *Neurosurg. Focus*, 2006, **21**, 1.
95. S. G. Keyrouz and M. N. Diringer, *Crit. Care*, 2007, **11**, 1.
96. M. Afzali, P. Ghaeli, M. Khanavi, M. Parsa, H. Montazeri, M. H. Ghahremani and S. N. Ostad, *Daru, J. Pharm. Sci.*, 2015, **23**, 1.
97. S. Sajadian, M. Vatankhah, M. Majdzadeh, S. M. Kouhsari, M. H. Ghahremani and S. N. Ostad, *Toxicol. Mech. Methods*, 2015, **25**, 388.
98. H. Huang, L.-J. Li, H.-B. Zhang and A.-Y. Wei, *J. BUON*, 2017, **22**, 112.
99. T. Goto, H. Matsushima, Y. Kasuya, Y. Hosaka, T. Kitamura, K. Kawabe, A. Hida, Y. Ohta, T. Simizu and K. Takeda, *Int. J. Urol.*, 1999, **6**, 314.
100. T. Shimizu, Y. Ohta, H. Ozawa, H. Matsushima and K. Takeda, *Anticancer Res.*, 2000, **20**, 761.
101. S. K. Noureini and M. Wink, *Molecules*, 2014, **19**, 11846.
102. A. Gaber, W. F. Alsanie, D. N. Kumar, M. S. Refat and E. M. Saied, *Molecules*, 2020, **25**, 5447.
103. S. E. Inzucchi, R. M. Bergenstal, J. B. Buse, M. Diamant, E. Ferrannini, M. Nauck, A. L. Peters, A. Tsapas, R. Wender and D. R. Matthews, *Diabetes Care*, 2012, **35**, 1364.
104. D. M. Nathan, J. B. Buse, M. B. Davidson, E. Ferrannini, R. R. Holman, R. Sherwin and B. Zinman, *Diabetes Care*, 2009, **32**, 193.
105. H. W. Rodbard, P. S. Jellinger, J. A. Davidson, D. Einhorn, A. J. Garber, G. Grunberger, Y. Handelsman, E. S. Horton, H. Lebovitz and P. Levy, *Endocr. Pract.*, 2009, **15**, 540.
106. J. Hyun Kim, S. Ofori, R. T. Mertens, S. Parkin and S. G. Awuah, *ChemMedChem*, 2021, **16**, 3222.
107. V. Craig Jordan, *Nat. Rev. Drug Discovery*, 2003, **2**, 205.
108. V. Scalcon, R. Bonsignore, J. Aupič, S. R. Thomas, A. Folda, A. A. Heidecker, A. Pöthig, A. Magistrato, A. Casini and M. P. Rigobello, *J. Med. Chem.*, 2023, **66**, 9823.

CHAPTER 4

Gold Complexes as Antibacterial Agents

The investigation of metal complexes with antibacterial properties is important in view of the increasing resistance of bacteria towards the available antibacterial drugs.[1] The initial development of antibacterial gold complexes was triggered by the findings of Robert Koch, who reported the antibacterial activity of potassium dicyanidoaurate(I), $K[Au(CN)_2]$ against *Mycobacterium tuberculosis*, which causes tuberculosis.[2] This discovery was followed by the design and development of a large reservoir of antibacterial gold(I) and gold(III) complexes with a diversity of ligand architectures.[3–5]

Currently, drug-resistant *Staphylococcus aureus* strains pose serious health threats to human populations worldwide. Biofilm formation of *S. aureus* shows a drug-tolerant nature with non-specific resistance towards several antibiotics. Metal complexes of N-heterocyclic carbene (NHC)-based ligands have shown promising antibacterial properties against several drug-resistant bacterial strains.[6] Ligands based on NHCs can be tuned to help in optimizing the properties of their complexes. Several gold-NHC complexes have been reported with interesting antibacterial properties.[7–9] Büssing *et al.*[10] reported antibacterial properties and inhibition of bacterial TrxR for a series of gold(I) and gold(III) N-heterocyclic monocarbene complexes (Figure 4.1, complexes **64–71**). The complexes displayed better antibacterial properties towards Gram-positive methicillin-resistant *S. aureus* (MRSA) and *Enterococcus faecium* strains than several Gram-negative bacteria, such as *Escherichia coli*, *Pseudomonas aeruginosa*, *Acinetobacter baumannii* and *Klebsiella pneumoniae*. In addition, the complexes efficiently inhibited TrxR, which indicated inhibition of this enzyme as the mechanism of action of their antibacterial effects. The research group of Palomero[11] reported antibacterial properties of a thiamine-inspired gold(I)-NHC complex (Figure 4.2, complex **72**) towards

The Therapeutic Potential of Gold Complexes
By Waseem A. Wani, Mohamed F. Alajmi, Afzal Hussain and Irshad Ahmad Wani

Published by the Royal Society of Chemistry, www.rsc.org

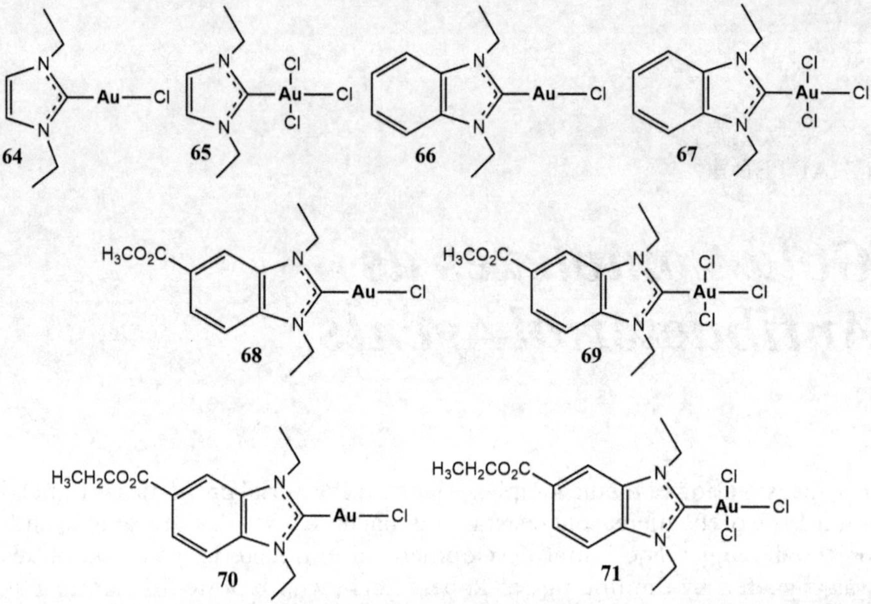

Figure 4.1 Chemical structures of gold(I) and gold(III) N-heterocyclic monocarbene complexes (**64–71**).[10]

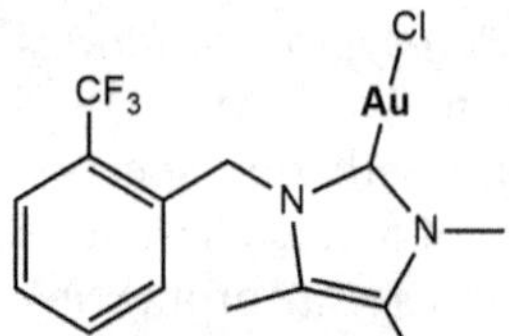

72

Figure 4.2 Chemical structure of thiamine-inspired gold(I)-NHC complex (**72**).[11]

S. aureus, *A. baumannii* and *P. aeruginosa*. The complex exhibited promising antibacterial activity against Gram-positive MRSA. However, the complex was less active towards Gram-negative *A. baumannii* and *P. aeruginosa*. The complex was more active than the standard antibacterial drug, silver sulfadiazine, and was equally effective as silver nitrate. This research is encouraging for the rational design and development of more active, broad-spectrum and biomimetically relevant gold-NHC complexes as antibacterial agents against a plethora of multidrug-resistant bacterial strains. Chen and co-workers[12] reported two gold(I) selenium-NHC complexes (Figure 4.3, complexes **73** and **74**) with better antibacterial properties towards several strains of multidrug-resistant (MDR) bacteria than the reference drug, auranofin. The complexes irreversibly inhibited the TrxR activity of bacteria by targeting the redox-active motif, thereby stopping the ability of TrxR to quench ROS, which finally led to oxidative stress. Recently, Rojo-Gómez *et al.*[13] documented the

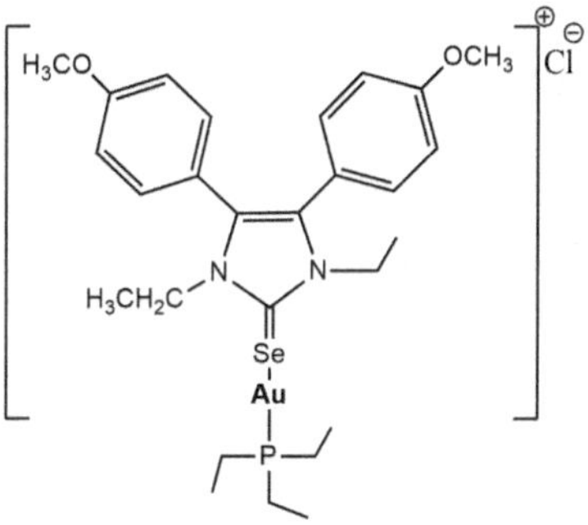
73

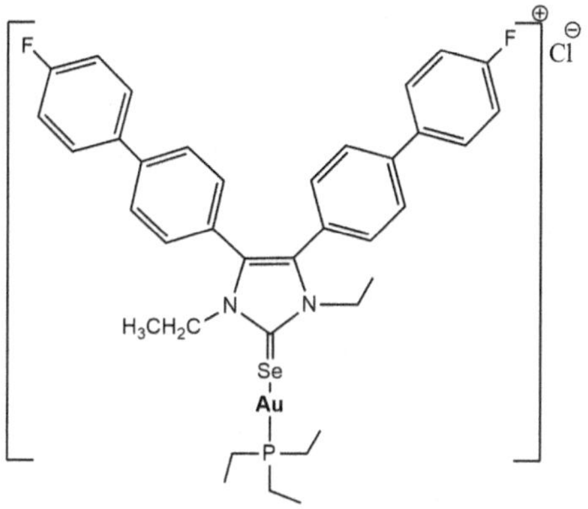
74

Figure 4.3 Chemical structures of gold(I) selenium-NHC complexes (**73** and **74**).[12]

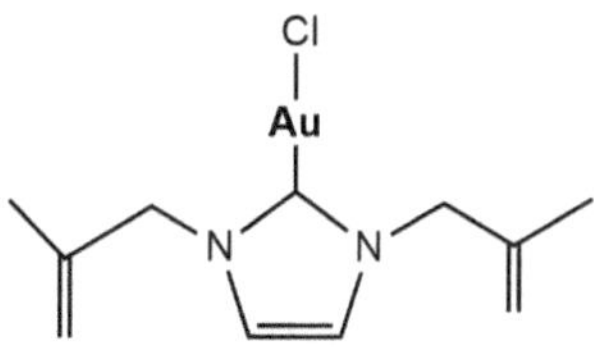
75

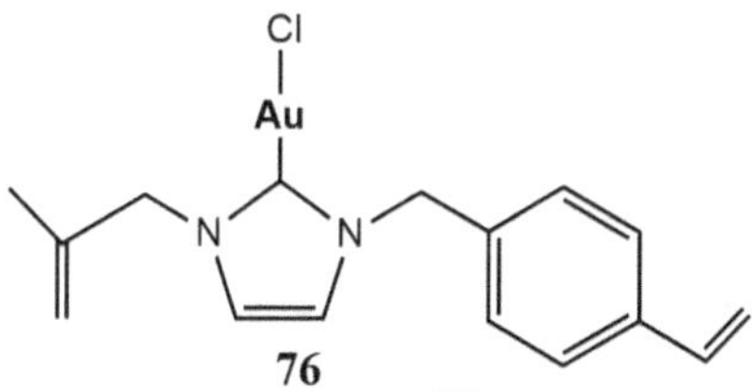
76

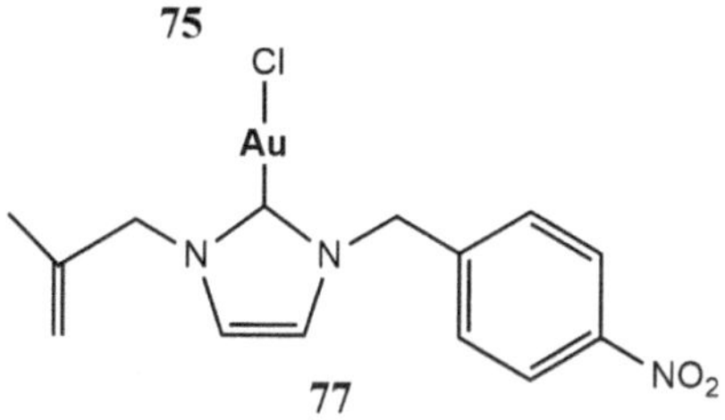
77

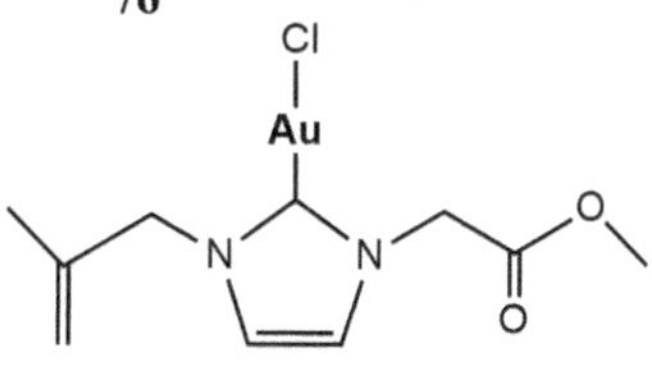
78

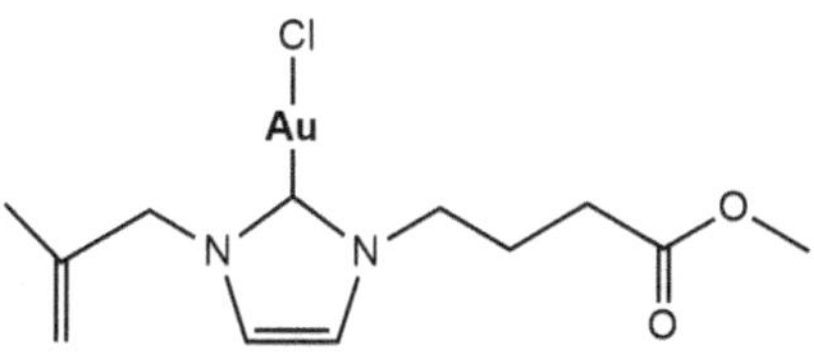
79

Figure 4.4 Chemical structures of functionalized monocarbene NHC-gold(I) complexes (**75–79**).[13]

antibacterial activities of a series of functionalized monocarbene NHC-gold(I) complexes (Figure 4.4, complexes **75–79**) against three Gram-positive bacterial strains, *S. aureus*, *Bacillus subtilis* and *Enterococcus faecium*, and three Gram-negative strains, *P. aeruginosa*, *E. coli* and *S. typhi*. All the complexes showed remarkable activity against the tested bacteria. Complexes **76** and **79** were highly active against *E. faecalis*. Interestingly, a very promising antibacterial activity was reported for complex **75** against *S. aureus* with the maximum inhibitory zone of 26 ± 0 mm. This report reveals the huge potential of functionalized monocarbene NHC-gold(I) complexes for the development of effective antibacterial agents.

Microorganisms have been known to display enhanced resistance to antimicrobial drugs in biofilm environments.[14] Formation of biofilms *in vivo* by *staphylococci* is a critical virulence factor for the perseverance of bacteria during infections.[15] In view of this, Mizdal *et al.*[6] documented the anti-biofilm properties of gold-complexed sulfonamides (Figure 4.5, complexes **80–84**) against *S. aureus* strains, including MRSA and clinical isolates. The gold-complexed sulfonamides remarkably reduced bacterial adhesion and biofilm formation. In addition, the complexes displayed potent antibacterial activities against *S. aureus* strains. Interestingly, the complexes displayed

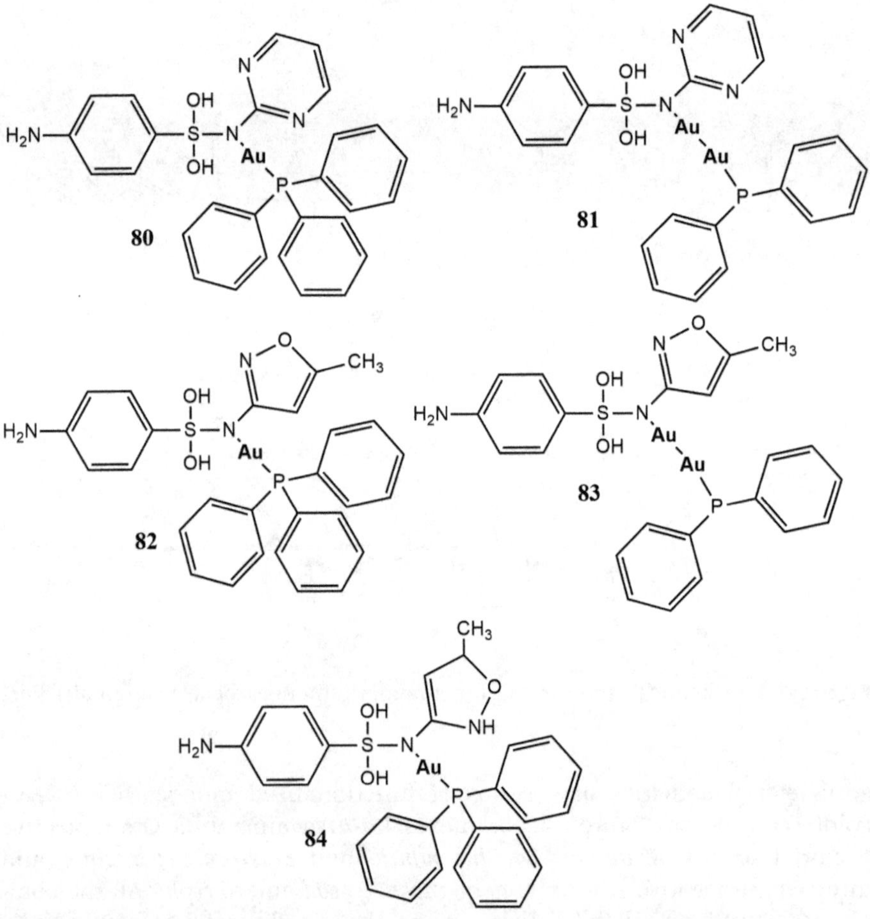

Figure 4.5 Chemical structures of gold-complexed sulfonamides (**80–84**).[6]

synergistic antibacterial activity in combination with classical antibiotics, such as ciprofloxacin and rifampicin.

Gold complexes of dithiocarbamate ligands have shown interesting biological properties over the years.[16] Ratia and co-workers[17] reported antibacterial properties of a gold(III)-dithiocarbamate complex (Figure 4.6, complex **85**) against several strains of Gram-positive and Gram-negative bacteria. In particular, the complex displayed promising antibacterial activity against Gram-positive strains. It is interesting to note that, after exposure of bacteria to the complex, no resistant mutants were detected. This indicated that the complex had a low tendency for developing resistance in the tested bacterial strains. The complex exerted its antibacterial action *via* a multimodal mechanism. The complex induced a strong reversible inhibition of TrxR. Moreover, the complex displayed low cytotoxicity in mammalian cell lines with no acute *in vivo* toxicity in mice. Abas *et al.*[18] reported antibacterial properties of several gold-dithiocarbamate complexes (Figure 4.7, complexes **86–91**)

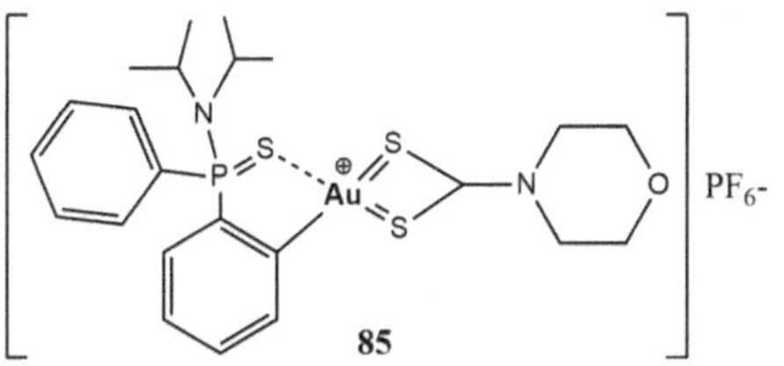

Figure 4.6 Chemical structure of gold(III)-dithiocarbamate complex (**85**).[17]

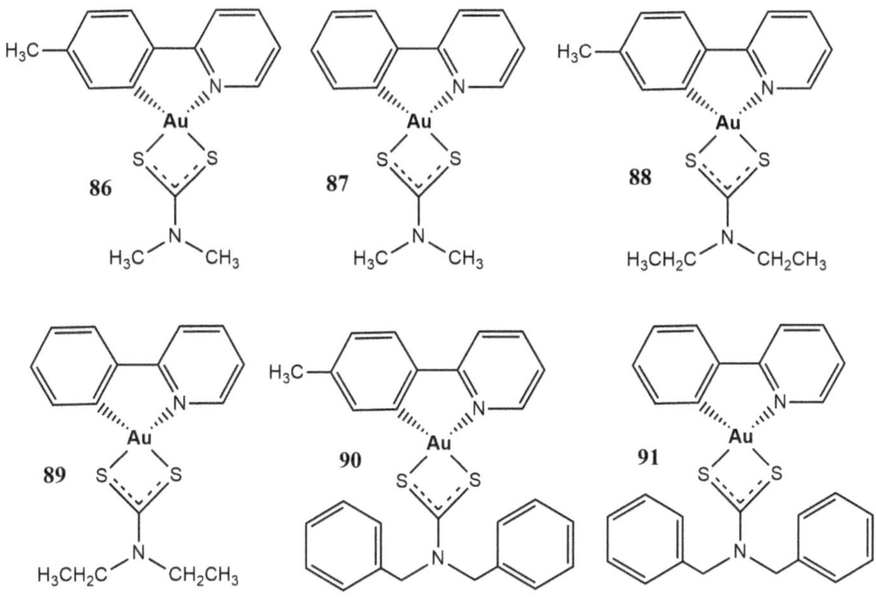

Figure 4.7 Chemical structures of gold-dithiocarbamate complexes (**86–91**).[18]

against *S. aureus*, *E. coli*, *K. pneumonia*, *P. aeruginosa* and *A. baumannii*. All the complexes were active against the tested bacterial strains. Complexes **86–89** were active against MRSA, with a minimum inhibitory concentration (MIC) of 0.25 μg mL^{-1}. In addition, complexes **86–89** were also effective towards *E. coli* (MIC = 32 μg mL^{-1}). Overall, this research report indicated that the complexes had better antibacterial activities, with more efficacy, against Gram-positive strains compared to Gram-negative ones.

Stability of gold complexes in biological media is one of the primary requirements for the design of therapeutic gold complexes. Gold(III) complexes of cyclometalated derivatives bearing C^N pincer-type ligands have demonstrated stability in biological media.[19] In this regard, Pintus *et al.*[20] reported the inhibition of biofilm formation by a heteroleptic cyclometalated gold(III) complex (Figure 4.8, complex **92**) bearing C-deprotonated 2-benzylpyridine and 1,2-dicyanoethene-1,2-dithiolate bidentate ligands. The complex displayed a significant bacteriostatic effect against *staphylococci*, with MIC values of 1.56 and 3.13 μg mL^{-1} against *Staphylococcus haemoliticus* and *S. aureus*, respectively. Ratia *et al.*[21] reported the antibacterial efficacy of a gold(III) complex, **93**, stabilized as C^S cycloaurated containing a diphenylphosphinothioic amide moiety (Figure 4.9). The complex displayed activity against Gram-positive and Gram-negative MDR bacterial strains, with MIC values in the range 4–8 and 6–32 mg L^{-1}, respectively. Interestingly, the complex displayed a biofilm inhibition of only two to four times that of its MIC. In addition, the complex exhibited a strong antibacterial synergistic effect in less than 2 h in combination with colistin. Also, the complex had a low

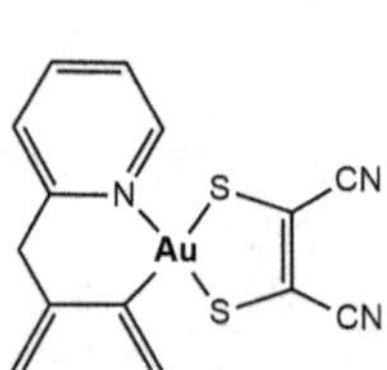

92

Figure 4.8 Chemical structure of heteroleptic cyclometalated gold(III) complex (**92**).[20]

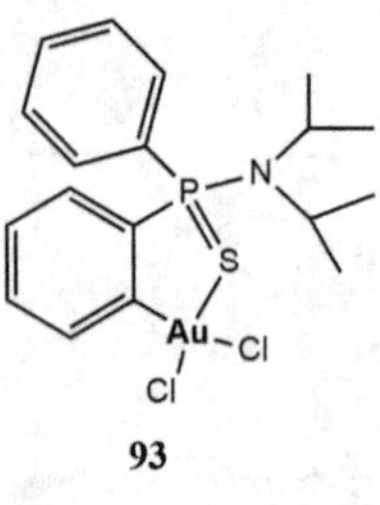

93

Figure 4.9 Chemical structure of gold(III) complex **93** stabilized as C^S-cycloaurated containing a diphenylphosphinothioic amide moiety.[21]

internalization rate in *S. aureus* and *A. baumannii*, and it did not interact with replication enzymes or efflux pumps. Importantly, there was no generation of resistant mutants in 30 day sequential cultures upon treatment with the complex. Moreover, lower cytotoxicity in the transformed human liver epithelial-2 (THLE-2) cell line, along with no signs of acute toxicity *in vivo* after an intravenous dose of 1 mg kg^{-1}, were reported. This research highlighted the reported complex as a potential future antibacterial agent. Besciani and co-workers[22] reported the antibacterial properties of an air- and water-stable gold-carbamate complex (Figure 4.10, complex **94**) against four pathogenic bacterial strains, including Gram-positive (vancomycin-resistant *E. faecium* and MRSA) and Gram-negative (*A. baumannii* and *P. aeruginosa*) strains. The complex displayed potent antibacterial properties against *E. faecium* (MIC = 5.0 μM) and MRSA (MIC = 3.1 μM), which was much better than for the reference drugs, silver nitrate and norfloxacin. This research report indicates that further investigation of air- and water-stable gold-carbamate complexes as potential antibacterial agents is warranted.

5-Fluorouracil is an established anticancer drug and is used for treating several types of cancer.[23] In addition to the antiproliferative effects of 5-fluorouracil, it has been reported to possess antibacterial properties against several bacterial strains.[24,25] The antibacterial effects of 5-fluorouracil are attributed to its potential to block virulence factors and biofilm production, in addition to stopping bacterial growth.[26–28] Owing to the therapeutic importance of 5-fluorouracil, Ferrando *et al.*[29] reported the antibacterial properties of phosphine-gold(I) complexes of 5-fluorouracil (Figure 4.11, complexes **95–99**) against *B. subtilis* and *E. coli*. All the complexes displayed better antibacterial properties compared to 5-fluorouracil. In addition, the complexes showed exciting antibacterial activity towards Gram-positive *B. subtilis* and were also active against Gram-negative *E. coli*. Interestingly, complex **99** containing the smallest phosphane moiety showed 32 times better activity than 5-fluorouracil. This research report revealed the enhancement of the antibacterial properties of 5-fluorouracil upon combination with a metallic center. This research encourages the approach of complexing therapeutically active molecules for the development of effective antibacterial agents.

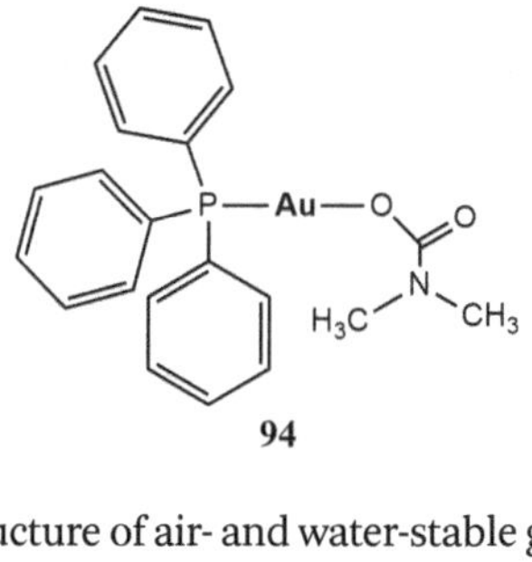

94

Figure 4.10 Chemical structure of air- and water-stable gold-carbamate complex **94**.[22]

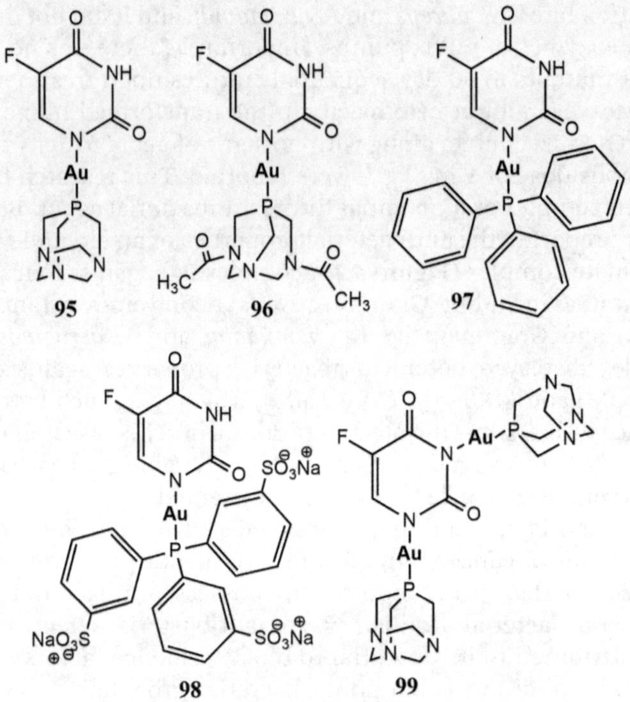

95 **96** **97** **98** **99**

Figure 4.11 Chemical structures of phosphine-gold(I) complexes of 5-fluorouracil (**95–99**).[29]

In summary, the research discussed here indicates that several types of structurally diverse and therapeutically important ligands/ligand systems have been employed by researchers for the development of antibacterial gold complexes. The antibacterial complexes are based on a gold center either in a +1 or a +3 oxidation state. The ligand systems used include NHCs, thiamine-inspired NHCs, selenium NHCs, functionalized monocarbene NHCs, dithiocarbamates, heteroleptic cyclometalated ligands and 5-fluorouracil. Some of the complexes have shown exciting and potent antibacterial activities towards several strains of pathogenic bacteria, including MRSA, *E. faecium*, *E. coli*, *P. aeruginosa*, *A. baumannii*, *K. pneumoniae*, *S. typhi* and *B. subtilis*. In addition, several complexes have shown better antibacterial activities than the market-available antimicrobials, such as silver sulfadiazine, silver nitrate, auranofin and ciprofloxacin. Interestingly, some complexes have displayed synergism of antibacterial effects when used in combination with standard antibacterial agents, such as ciprofloxacin, norfloxacin, colistin and rifampicin. The use of several complexes has also been used to stop the generation of mutant-resistant strains of bacteria. Overall, the gold complexes discussed in this chapter provide clear evidence of the enormous potential of these complexes as candidates for the design and development of antibacterial agents.

References

1. A. Hussain, M. F. Alajmi, M. A. Lone and W. A. Wani, *Therapeutic Rhodium Complexes*, Springer Nature Switzerland, Cham, 2023, pp. 13–53.
2. G. J. Higby, *Gold Bull.*, 1982, **15**, 130–140.
3. B. \DJ Glišić and M. I. Djuran, *Dalton Trans.*, 2014, **43**, 5950–5969.
4. C. Ratia, S. Sueiro, R. G. Soengas, M. J. Iglesias, F. López-Ortiz and S. M. Soto, *Antibiotics*, 2022, **11**, 1728.
5. C. Ratia, R. G. Soengas and S. M. Soto, *Front. Microbiol.*, 2022, **13**, 846959.
6. C. R. Mizdal, S. T. Stefanello, V. da Costa Flores, V. A. Agertt, P. C. Bonez, G. G. Rossi, T. C. da Silva, F. A. A. Soares, L. de Lourenço Marques and M. M. A. de Campos, *Microb. Pathog.*, 2018, **123**, 440–448.
7. C. Schmidt, B. Karge, R. Misgeld, A. Prokop, R. Franke, M. Brönstrup and I. Ott, *Chem. - Eur. J.*, 2017, **23**, 1869–1880.
8. C. Schmidt, L. Albrecht, S. Balasupramaniam, R. Misgeld, B. Karge, M. Brönstrup, A. Prokop, K. Baumann, S. Reichl and I. Ott, *Metallomics*, 2019, **11**, 533–545.
9. C. Schmidt, B. Karge, R. Misgeld, A. Prokop, M. Brönstrup and I. Ott, *MedChemComm*, 2017, **8**, 1681–1689.
10. R. Büssing, B. Karge, P. Lippmann, P. G. Jones, M. Brönstrup and I. Ott, *ChemMedChem*, 2021, **16**, 3402–3409.
11. O. E. Palomero, A. L. Cunningham, B. W. Davies and R. A. Jones, *Inorg. Chim. Acta*, 2021, **517**, 120152.
12. X. Chen, S. Sun, S. Huang, H. Yang, Q. Ye, L. Lv, Y. Liang, J. Shan, J. Xu and W. Liu, *Redox Biol.*, 2023, **60**, 102621.
13. E. G. Rojo-Gómez, F. J. Zuno-Cruz, G. Sánchez-Cabrera, M. Carpio-Granillo, D. O. González-Ábrego, C. Coronel-Olivares, J. G. Alvarado-Rodríguez and M. J. Rosales-Hoz, *Inorg. Chim. Acta*, 2023, **555**, 121557.
14. A. Algburi, N. Comito, D. Kashtanov, L. M. T. Dicks and M. L. Chikindas, *Appl. Environ. Microbiol.*, 2017, **83**, e02508-16.
15. J. L. Lister and A. R. Horswill, *Front. Cell. Infect. Microbiol.*, 2014, **4**, 178.
16. C. I. Yeo, E. R. Tiekink and J. Chew, *Inorganics*, 2021, **9**, 48.
17. C. Ratia, V. Ballén, Y. Gabasa, R. G. Soengas, M. Velasco-de Andrés, M. J. Iglesias, Q. Cheng, F. Lozano, E. S. Arnér and F. López-Ortiz, *Front. Microbiol.*, 2023, **14**, 1198473.
18. E. Abás, D. Aguirre-Ramírez, M. Laguna and L. Grasa, *Biomedicines*, 2021, **9**, 1775.
19. P. Chakraborty, D. Oosterhuis, R. Bonsignore, A. Casini, P. Olinga and D. Scheffers, *ChemMedChem*, 2021, **16**, 3060–3070.
20. A. Pintus, M. C. Aragoni, M. A. Cinellu, L. Maiore, F. Isaia, V. Lippolis, G. Orrù, E. Tuveri, A. Zucca and M. Arca, *J. Inorg. Biochem.*, 2017, **170**, 188–194.
21. C. Ratia, V. Cepas, R. Soengas, Y. Navarro, M. Velasco-de Andrés, M. J. Iglesias, F. Lozano, F. López-Ortiz and S. M. Soto, *Front. Microbiol.*, 2022, **13**, 815622.
22. G. Bresciani, N. Busto, V. Ceccherini, M. Bortoluzzi, G. Pampaloni, B. Garcia and F. Marchetti, *J. Inorg. Biochem.*, 2022, **227**, 111667.

23. C. Heidelberger, N. K. Chaudhuri, P. Danneberg, D. Mooren, L. Griesbach, R. Duschinsky, R. J. Schnitzer, E. Pleven and J. Scheiner, *Nature*, 1957, **179**, 663–666.
24. J. M. Thomson and I. L. Lamont, *Front. Microbiol.*, 2019, **10**, 952.
25. C. Oe, H. Hayashi, K. Hirata, K. Kawaji, F. Hashima, M. Sasano, M. Furuichi, E. Usui, M. Katsumi, Y. Suzuki, C. Nakajima, M. Kaku and E. N. Kodama, *ACS Infect. Dis.*, 2020, **6**, 1490–1500.
26. A. Rangel-Vega, L. R. Bernstein, E. A. Mandujano-Tinoco, S. J. García-Contreras and R. García-Contreras, *Front. Microbiol.*, 2015, **6**, 282.
27. A. Ueda, C. Attila, M. Whiteley and T. K. Wood, *Microb. Biotechnol.*, 2009, **2**, 62–74.
28. C. Attila, A. Ueda and T. K. Wood, *Appl. Microbiol. Biotechnol.*, 2009, **82**, 525–533.
29. R. Ferrando, S. G. Mitchell, E. Atrián-Blasco and E. Cerrada, *Dalton Trans.*, 2023, **52**, 10507–10514.

CHAPTER 5

Gold Complexes as Antifungal Agents

The antifungal mechanisms of action of metal complexes are completely metal specific, which hinders the development of resistance mechanisms in pathogenic fungi.[1] Metal complexes of several transition metals with promising antifungal activities have been reported.[2–5] Gold complexes with different types of ligand have been shown to display exciting antifungal properties against a plethora of disease-causing fungi.[6] In 2000, Nomiya and co-workers[7] reported the antifungal properties of gold(I)-triphenylphosphine complexes with nitrogen-containing heterocycles (Figure 5.1, complexes **100** and **101**). The complexes exhibited moderate antifungal properties against *Candida albicans*, with MIC values ranging from 31.3 to more than 1000 μg mL^{-1}. Denis *et al.*[8] documented the antifungal activities of gold(I)-phosphine complexes (Figure 5.2, complexes **102–107**) towards 20 fungal strains, including seven *Candida* spp., five non-*Candida albicans* spp., four *Cryptococcus* spp., three *Aspergillus* spp. and one *Fusarium* spp. Complexes **105** and **107** exhibited outstanding activity towards *Candida* spp. and *Cryptococcus* spp., with MIC values in the range 0.06–1.95 μg mL^{-1}, and these values were better than the values for amphotericin B. Complexes **105** and **107** also exhibited promising activity, with MIC values in the range 1.95–7.8 μg mL^{-1} against the filamentous fungi, namely, *Aspergillus* spp. and *Fusarium* spp. These MIC values were much better than those for amphotericin B. This research indicated that the linear gold(I) complexes (**102, 104** and **106**) and achiral square-planar gold(I) complex (**103**) were weak antifungals that should not be investigated further. However, the chiral square-planar gold(I) complexes (**105** and **107**) displayed great promise as antifungal agents and so deserve additional investigation.

The Therapeutic Potential of Gold Complexes
By Waseem A. Wani, Mohamed F. Alajmi, Afzal Hussain and Irshad Ahmad Wani

Published by the Royal Society of Chemistry, www.rsc.org

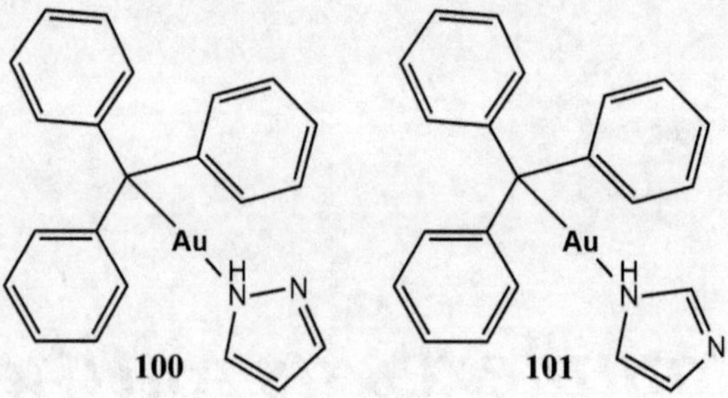

Figure 5.1 Chemical structures of gold(I)-triphenylphosphine complexes with nitrogen-containing heterocycles (**100** and **101**).[7]

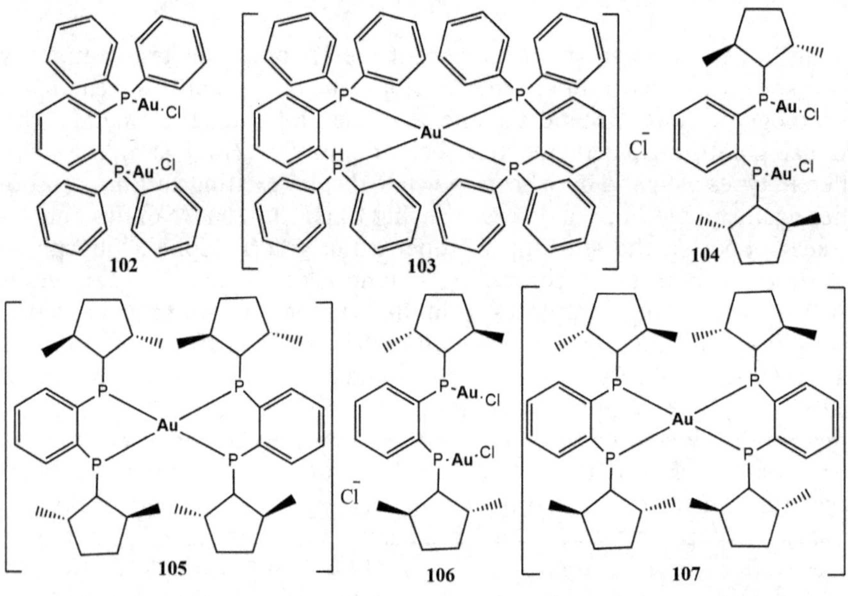

Figure 5.2 Chemical structures of gold(I)-phosphine complexes (**102–107**).[8]

N-Heterocyclic carbene (NHC) complexes of several transition metal ions, including gold, have been thoroughly investigated as antifungal agents.[9,10] NHCs complexed with gold ions have displayed interesting activities against a plethora of fungal strains.[11] Ozdemir *et al.*[12] reported the antifungal activities of gold(I) complexes of NHCs (Figure 5.3, complexes **108–110**) against two pathogenic *Candida* species, *C. albicans* and *C. tropicalis*. The complexes were active against the tested fungal strains, with MIC values in the range 12.5–200 μg mL^{-1}. Complex **109** was the most active complex, with an MIC value of 12.5 μg mL^{-1} against both *C. albicans* and *C. tropicalis*. The authors extended this work and documented the antifungal activities of anionic gold-NHC complexes (Figure 5.4, complexes **111–116**) against *C. albicans* and

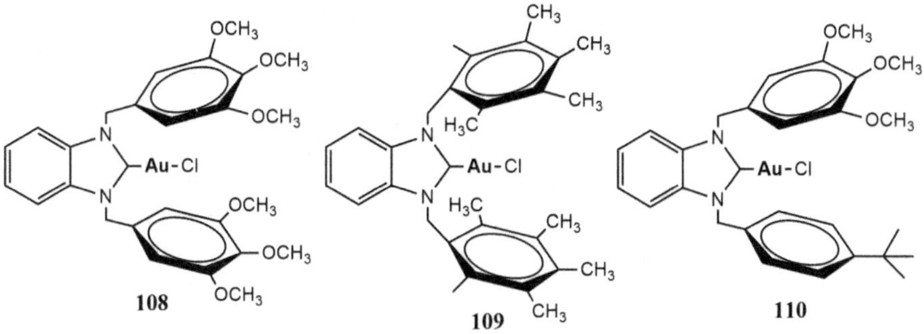

Figure 5.3 Chemical structures of gold(I)-NHC complexes (**108–110**).[12]

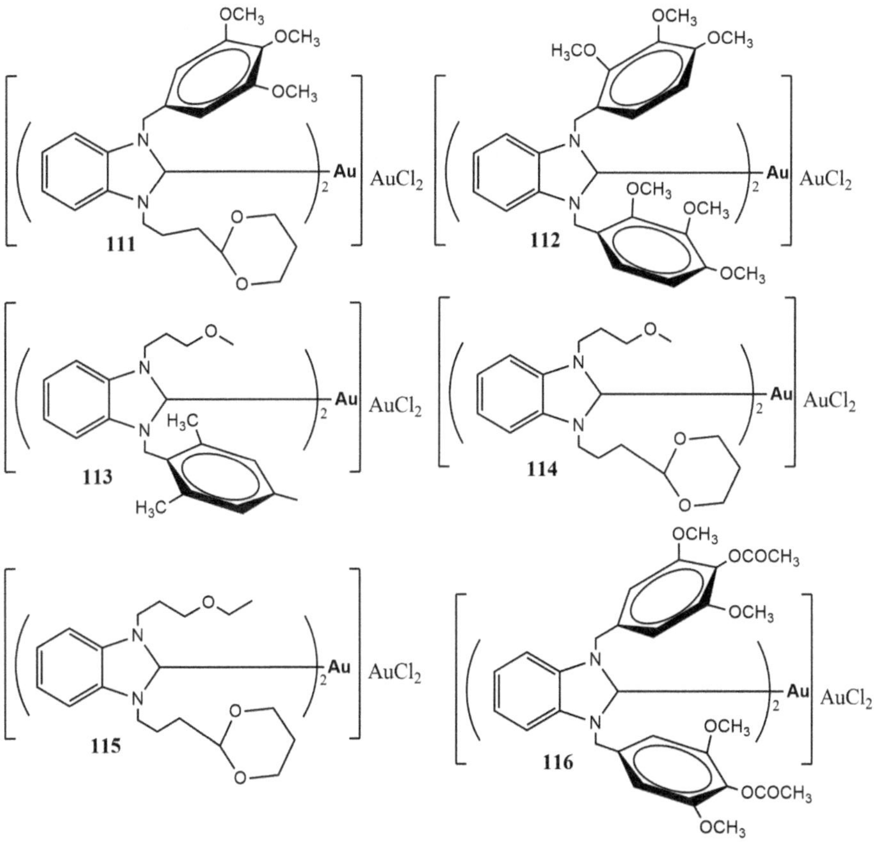

Figure 5.4 Chemical structures of anionic gold-NHC complexes (**111–116**).[13]

C. tropicalis. The complexes were active against the tested fungal strains, with MIC values in the range 12.5–100 μg mL^{-1}. It is interesting to note that each complex displayed a similar activity profile against both fungal strains. The order of activity of the complexes towards the two fungal strains was **112** = **113** > **115** > **114** = **116** > **111**.[13]

Recently, metal–drug chelation chemistry has been used in the design of drugs for enhancement of biological properties.[14] The metal ions, on complexation with the drug, play a critical role in modifying the toxicological and pharmacological properties of that drug.[15,16] Significant enhancement in the medical efficacies of several drugs have been reported with some metal ions specifically in living systems.[17] Chloramphenicol is a bacteriostatic agent and is widely used for the treatment of ophthalmic bacterial infections.[18] Making use of the metal–drug chelation approach, Al-Khodir *et al.*[19] documented the antifungal activity of an Au(III) complex of chloramphenicol (Figure 5.5, complex **117**) against *Aspergillus flavus* and *C. albicans*. The complex was active against both *A. flavus* and *C. albicans*, with inhibition zone diameters of 5 mm and 7 mm, respectively. It is interesting to note that the complex was much more active than free chloramphenicol. Gagni *et al.*[20] reported antifungal properties of two gold complexes (Figure 5.6, complexes

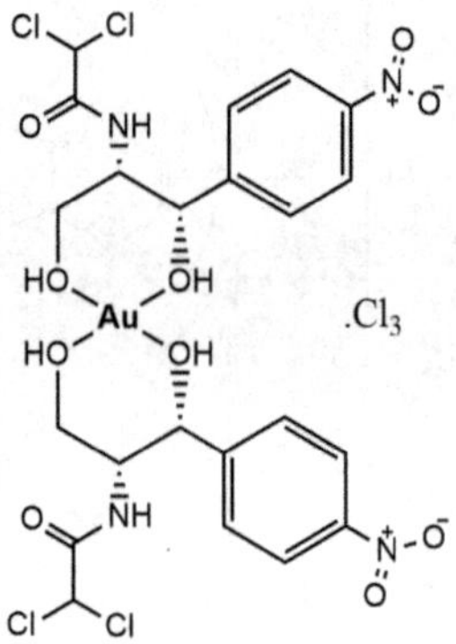

117

Figure 5.5 Chemical structure of Au(III) complex of chloramphenicol (**117**).[19]

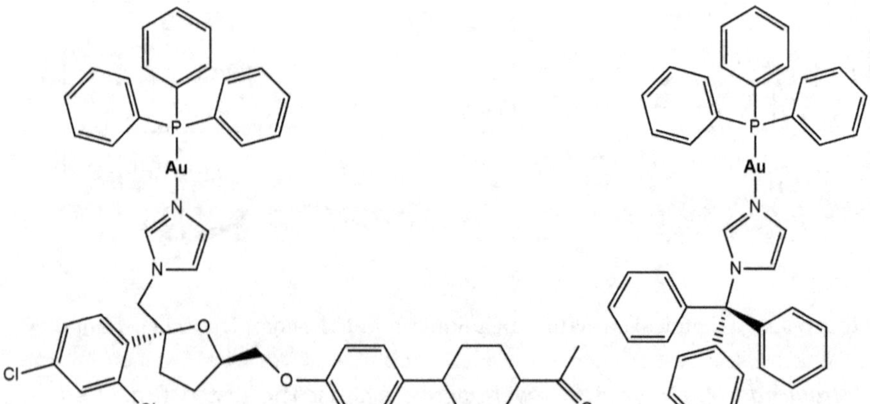

118 **119**

Figure 5.6 Chemical structures gold complexes (**118** and **119**) bearing antifungal ligands, ketoconazole and clotrimazole.[20]

118 and **119**) bearing the antifungal ligands ketoconazole and clotrimazole against *sporotrichosis* agents (*S. schenckii*, *S. brasiliensis* and *S. globosa*). The complexes inhibited the growth of fungi, as well as killing them at nanomolar concentrations. Interestingly, the complexes were more potent than clotrimazole and ketoconazole alone. Moreover, the complexes displayed better selectivity towards fungi than mammalian cells and human red blood cells. This research highlights the potential for gold complexes bearing antifungal ligands as promising leads in the design and development of potent antifungal agents. Recently, Stevanovic and co-workers[21] employed several clinically used antifungal azoles (clotrimazole, econazole, tioconazole and voriconazole) as ligands for the development of gold(III) antifungal agents. The gold(III) complexes (Figure 5.7, complexes **120–123**) were screened against *C. albicans*, *C. parapsilosis*, *C. glabrata*, *C. krusei* and *C. auris*. These complexes, containing the clinically used antifungal drugs, displayed amplified antifungal efficacies in comparison to their azole ligands. Maximum enhancement in activity was obtained with the complexes containing tioconazole and voriconazole as ligands. This research report highlights a powerful approach for the development of gold(III)-based antivirulence antifungal drugs with the ability to reduce virulence without a marked effect on pathogen growth, which offers a lower risk for the development of resistance mechanisms in the pathogenic fungi.

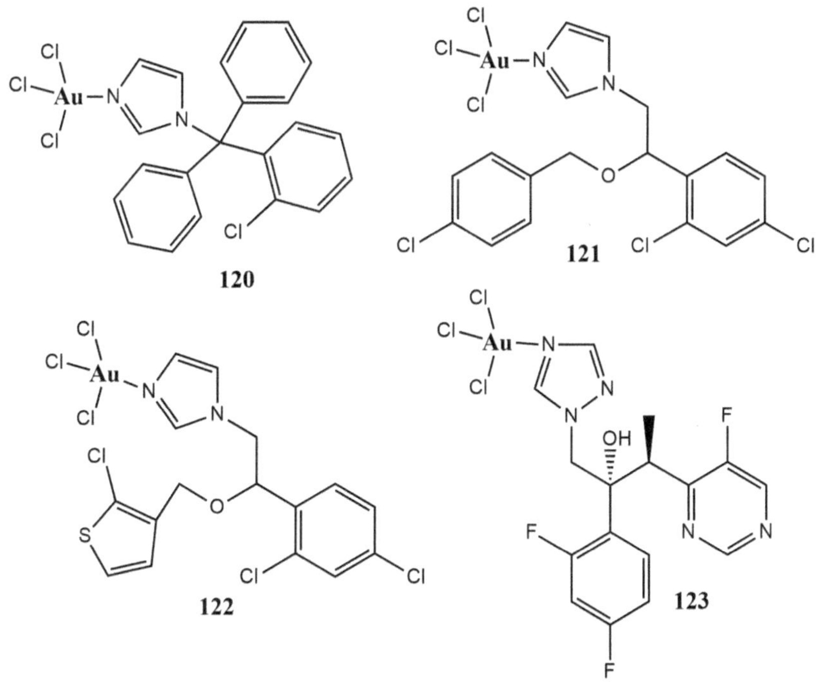

Figure 5.7 Chemical structures of gold(III) complexes with clinically used antifungal agents as ligands (**120–123**).[21]

Metal complexes of benzimidazole ligands have gained fame owing to their exciting biological properties, such as anticancer,[22] antimicrobial,[23] anti-helminthic[24] and DNA intercalation.[25] In addition, benzimidazole ligands have been employed in the design of frameworks for modeling biological systems as a result of their strong coordinating potential.[26] Mansour and Shehab[27] reported interesting antifungal activities of three pyridylbenzimidazole-based gold(III) complexes (Figure 5.8, complexes **124–126**) against *C. albicans* and *Cryptococcus neoformans* var. *grubii*. The complexes displayed promising antifungal activities against the tested fungal strains, with MIC values in the range 0.25–32 μg mL^{-1}. The order of activity of the complexes was **124** > **126** > **125**. Complex **124** was the most active complex, with MIC values of 0.25 μg mL^{-1} against both fungal strains.

Metal complexes of thiourea-based ligands have shown interesting bioactivities, including DNA-binding interactions, antimicrobial activities, anticancer properties, anti-arthritic efficacy and antioxidant properties.[28–30] Khan *et al.*[31] reported antifungal properties of a gold(I) complex of 2,4,6-trimethylphenyl-3-benzoylthiourea (Figure 5.9, complex **127**) against

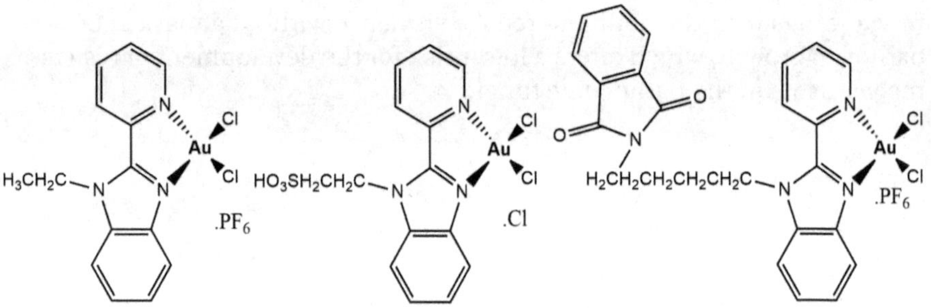

124 125 126

Figure 5.8 Chemical structures of pyridylbenzimidazole-based gold(III) complexes (**124–126**).[27]

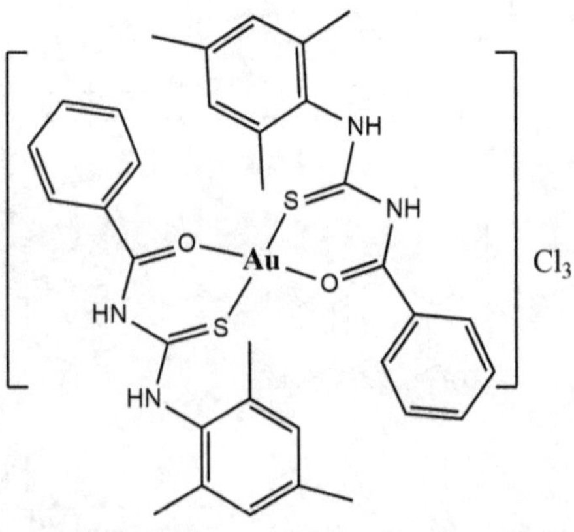

127

Figure 5.9 Chemical structure of gold(I) complex of 2,4,6-trimethylphenyl-3-benzoylthiourea (**127**).[31]

three fungal strains, namely, *Aspergillus fumigatus*, *Fusarium solani* and *Mucor* species. The complex was active against *A. fumigatus* and *F. solani*, with zones of inhibition of 7.5 ± 0.707 mm and 17.5 ± 1.11 mm, respectively. However, the complex was inactive towards *Mucor* species. Importantly, the complex was less active than the standard drug, terbinafine, against the tested fungal strains.

A critical analysis of the research described in this chapter reveals that several biologically important ligands have been complexed with gold metal ions for the development of gold(I) and gold(III) complexes with exciting antifungal properties. A variety of ligand systems, including triphenylphosphine, nitrogen-containing heterocycles, NHCs, bacteriostatic drugs (*e.g.* chloramphenicol), clinically used antifungal azoles (*e.g.* imidazole, 1-isopropylimidazole, 1-phenylimidazole, clotrimazole, econazole, tioconazole and voriconazole), ketoconazole, benzimidazole and thiourea derivatives, have been used for the design and development of antifungal gold complexes. Some complexes have exhibited excellent antifungal activities against pathogenic fungi, including *Cryptococcus* spp., *Fusarium* spp., *C. albicans*, *C. tropicalis*, *A. flavus*, *C. parapsilosis*, *C. glabrata*, *C. krusei* and *C. auris*. Importantly, some complexes have exhibited much better antifungal properties compared to the standard antifungal drugs, such as amphotericin B, clotrimazole, econazole, tioconazole and voriconazole. In addition, some complexes have shown glimpses of behaving as antivirulence antifungal drugs. This may help to reduce virulence without an obvious effect on pathogen growth, which might offer a lower risk of the onset of resistance mechanisms in the pathogenic fungi. In a nutshell, the gold complexes outlined in this chapter deliver crystal clear evidence of the huge potential of gold complexes for the rational design and development of antifungal agents.

References

1. Y. Lin, H. Betts, S. Keller, K. Cariou and G. Gasser, *Chem. Soc. Rev.*, 2021, 10346–10402.
2. E. Rodriguez-Fernandez, J. L. Manzano, J. J. Benito, R. Hermosa, E. Monte and J. J. Criado, *J. Inorg. Biochem.*, 2005, **99**, 1558–1572.
3. Z. H. Chohan, M. Arif, M. A. Akhtar and C. T. Supuran, *Bioinorg. Chem. Appl.*, 2006, **2006**, 083131.
4. K. Bajaj, R. M. Buchanan and C. A. Grapperhaus, *J. Inorg. Biochem.*, 2021, **225**, 111620.
5. J. Kljun, A. J. Scott, T. Lanišnik Rižner, J. Keiser and I. Turel, *Organometallics*, 2014, **33**, 1594–1601.
6. B. D. Glišić and M. I. Djuran, *Dalton Trans.*, 2014, **43**, 5950–5969.
7. K. Nomiya, R. Noguchi, K. Ohsawa, K. Tsuda and M. Oda, *J. Inorg. Biochem.*, 2000, **78**, 363–370.
8. E. K. Dennis, J. H. Kim, S. Parkin, S. G. Awuah and S. Garneau-Tsodikova, *J. Med. Chem.*, 2020, **63**, 2455–2469.
9. L. Ronga, M. Varcamonti and D. Tesauro, *Molecules*, 2023, **28**, 4435.

10. A. Alsalhi, *J. Coord. Chem.*, 2023, **76**, 847–861.
11. S. Díez-González, *N-Heterocyclic Carbenes: From Laboratory to Curiosities to Efficient Synthetic Tools*, Royal Society of Chemistry, 2016, vol. 27.
12. İ. Özdemir, N. Temelli, S. Günal and S. Demir, *Molecules*, 2010, **15**, 2203–2210.
13. Ö. Doğan, N. Kaloğlu, S. Demir, İ. Özdemir, S. Günal and İ. Özdemir, *Monatsh. Chem.*, 2013, **144**, 313–319.
14. L. Tabrizi, H. Chiniforoshan and H. Tavakol, *Spectrochim. Acta, Part A*, 2015, **141**, 16–26.
15. L.-F. Li, H. Wang, J. Zhang, C. Ma, Y.-Y. Li, L. Wang, S.-K. Liang, H.-T. Jin, S.-J. Liu and M.-C. Zhu, *Eur. J. Med. Chem.*, 2015, **92**, 295–301.
16. Q.-P. Qin, Z.-F. Chen, J.-L. Qin, X.-J. He, Y.-L. Li, Y.-C. Liu, K.-B. Huang and H. Liang, *Eur. J. Med. Chem.*, 2015, **92**, 302–313.
17. N. L. G. D. de Souza, H. C. Garcia, M. C. de Souza, A. L. do Amaral Fernandes, G. C. Pereira, R. Diniz and L. F. C. de Oliveira, *J. Mol. Struct.*, 2015, **1085**, 21–27.
18. B. I. Bale, E. E. Elebesunu, P. Manikavasagar, F. O. Agwuna, I. O. Ogunkola, A. U. Sow and D. E. Lucero-Prisno, *Trop. Med. Health*, 2023, **51**, 15.
19. F. A. Al-Khodir and M. S. Refat, *J. Mol. Struct.*, 2016, **1119**, 157–166.
20. T. Gagini, L. Colina-Vegas, W. Villarreal, L. P. Borba-Santos, C. de Souza Pereira, A. A. Batista, M. K. Fleury, W. de Souza, S. Rozental and L. A. S. Costa, *New J. Chem.*, 2018, **42**, 13641–13650.
21. N. L. Stevanović, J. Kljun, I. Aleksic, S. S. Bogojevic, D. Milivojevic, A. Veselinovic, I. Turel, M. I. Djuran, J. Nikodinovic-Runic and B. D. Glišić, *Dalton Trans.*, 2022, **51**, 5322–5334.
22. F. Gümüş, G. Eren, L. Açık, A. Çelebi, F. Öztürk, Ş. Yılmaz, R. I. Sağkan, S. Gür, A. Özkul, A. Elmalı and Y. Elerman, *J. Med. Chem.*, 2009, **52**, 1345–1357.
23. N. T. Abdel-Ghani, M. F. A. El-Ghar and A. M. Mansour, *Spectrochim. Acta, Part A*, 2013, **104**, 134–142.
24. G. Yadav and S. Ganguly, *Eur. J. Med. Chem.*, 2015, **97**, 419–443.
25. V. Rajendiran, M. Murali, E. Suresh, S. Sinha, K. Somasundaram and M. Palaniandavar, *Dalton Trans.*, 2008, 148–163.
26. E. Bouwman, W. L. Driessen and J. Reedijk, *Coord. Chem. Rev.*, 1990, **104**, 143–172.
27. A. M. Mansour and O. R. Shehab, *Eur. J. Inorg. Chem.*, 2019, **2019**, 2830–2838.
28. A. Colak, Ü. Terzi, M. Col, Ş. A. Karaoglu, S. Karaböcek, A. Küçükdumlu and F. A. Ayaz, *Eur. J. Med. Chem.*, 2010, **45**, 5169–5175.
29. D. Li, Y. Sun, S. Liu and Z. Hua, *J. Coord. Chem.*, 2006, **59**, 403–408.
30. Z.-C. Liu, B.-D. Wang, B. Li, Q. Wang, Z.-Y. Yang, T.-R. Li and Y. Li, *Eur. J. Med. Chem.*, 2010, **45**, 5353–5361.
31. U. A. Khan, A. Badshah, M. N. Tahir and E. Khan, *Polyhedron*, 2020, **181**, 114485.

CHAPTER 6

Gold Complexes as Antiviral Agents

Investigations of the antiviral potential of gold complexes have been scarce but scientific interest in the development of gold complexes with antiviral properties has taken off over the last couple of decades. Antirheumatic gold(I) thiolates in general, and auranofin in particular, have been documented to possess interesting antiviral properties against human immunodeficiency virus (HIV), the causative agent of acquired immune deficiency syndrome (AIDS).[1,2] Aurothiolates are known to inhibit the proliferation of HIV *via* immunomodulation.[3] Auranofin has been reported to decrease the virus reservoir in monkeys infected with simian immunodeficiency virus. In addition, auranofin also reduces the viral load in patients infected with HIV.[4] Beyond aurothiolates, gold(I) complexes containing phosphorus-based ligands have exhibited exciting inhibitory activities against HIV. These complexes have been reported to cause direct inhibition of the HIV reverse transcriptase enzyme.[5] Recently, gold complexes with different ligand systems have been investigated against several viruses, and have shown promising antiviral properties.[6]

The worldwide pandemic due to HIV/AIDS started in 1981, and it is still an ongoing serious public health issue around the world. According to data from the World Health Organization (WHO), nearly 40.4 million people have died and around 39 million people have been infected as a result of HIV globally. In 2022, approximately 630 000 deaths worldwide occurred due to HIV/AIDS.[7] Owing to the damage caused by HIV to human populations across different regions of the world, the development of anti-HIV drugs is very important. In this regard, in 1993,[8] Okada *et al.* reported strong inhibition of HIV-1 infectivity through ligand exchange of bis(thiolato)gold(I) with a component of the surface of the virion. This research triggered the design and development of other gold complexes with structurally diverse ligands as antiviral agents.

The Therapeutic Potential of Gold Complexes
By Waseem A. Wani, Mohamed F. Alajmi, Afzal Hussain and Irshad Ahmad Wani

Published by the Royal Society of Chemistry, www.rsc.org

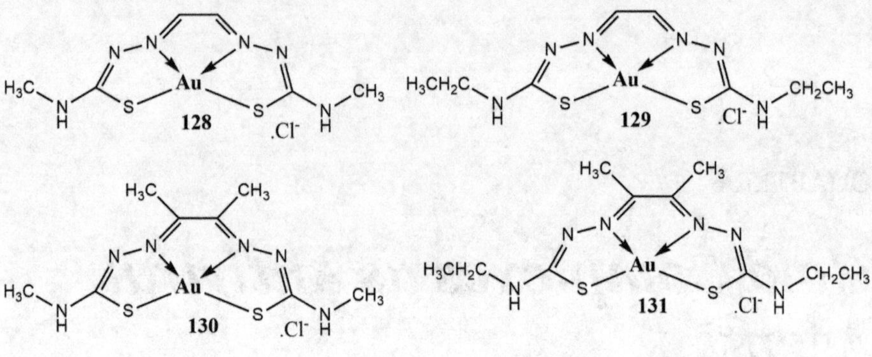

Figure 6.1 Chemical structures of gold(III) complexes of bis(thiosemicarbazones) (**128–131**).[9]

Fonteh and co-workers[9] documented the inhibition of the replication of HIV at cytostatic concentrations by using gold(III) complexes of bis(thiosemicarbazones) (Figure 6.1, complexes **128–131**). Complex **130** was reported to inhibit the viral infection of TZM-bl cells by 98%, with an IC_{50} value of 6.8 ± 0.6 µM at a non-toxic concentration of 12.5 µM. Complex **131** induced inhibition of infection of these cells by 72% and 98%, with an IC_{50} value of 5.3 ± 0.4 µM, at concentrations of 6.25 and 12.5 µM, respectively. Bjelosevic *et al.*[10] reported anti-HIV properties of gold(I) complexes of 1,10-bis(diphenylphosphino)metallocene derivatives (Figure 6.2, complexes **132–136**). Complex **136** exhibited strong viral inhibition, with an IC_{50} value of 0.5 ± 0.2 µM. The antiviral activity of the complex was far superior to that of the standard antiviral drug azidothymidine (IC_{50} = 1.3 ± 0.3 µM) in the CEM-SS (human T lymphoid) cell line. In addition, the selectivity index of the complex was around 12-fold higher than that of azidothymidine. This research indicated the strong potential of the complex as an anti-HIV agent. Gama and co-workers[11] demonstrated the virostatic potential of a gold(I) complex of 1,10-bis(diphenylphosphino)ferrocene–quinoline conjugate (Figure 6.3, complex **137**) against HIV-1. The complex displayed substantial inhibition of the infectivity of the virus (83%) at a non-toxic concentration of 10 µg mL^{-1}. The same concentration of the complex induced cytostatic behavior in TZM-bl cells, along with S and G2/M phase cell cycle arrest, which indicated the potential of the complex as a virostatic agent with both antiviral and cytostatic properties. This dual action with respect to antiviral and cytostatic effects was thought to be able to avoid the use of combination drugs with their consequent negative effects, including drug–drug interactions. Overall, this research strongly supports the further development of this complex for future applications as a virostatic drug in HIV-1 infection.

HIV latent reservoirs are a result of combination of the cDNA of the virus into the genome of the host cell.[12] Viral latent reservoirs lead to viral persistence even in the presence of, and after treatment with, effective antiviral regimens or immune retorts.[13–15] The cells that are latently infected can replicate competently after reactivation[16] and therefore increase the viral

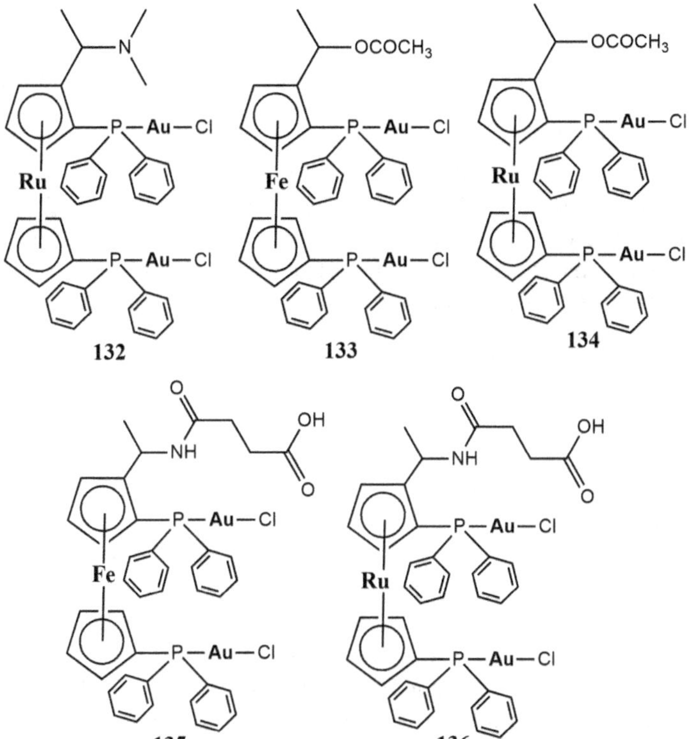

132 133 134 135 136

Figure 6.2 Chemical structures of gold(I) complexes of 1,10-bis(diphenylphosphino)metallocene derivatives (**132–136**).[10]

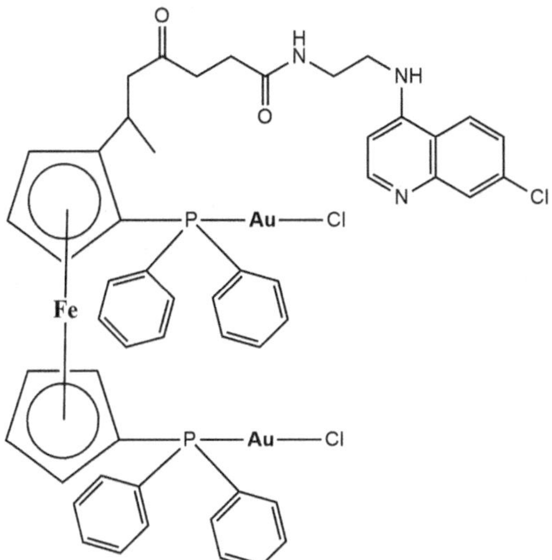

137

Figure 6.3 Chemical structure of the gold(I) complex of 1,10-bis(diphenylphosphino)ferrocene–quinoline conjugate (**137**).[11]

load even after antiretroviral therapy intervention.[17,18] Reactivation of latent viral reservoirs is considered to be a possible strategy for the elimination of latency and consequent HIV re-infection.[19] This activation/elimination strategy aims at flushing out the latent viruses by induction of the expression of certain viral proteins.[20] The production of these viral proteins indicates an increase in the viral cytopathic effect, which makes the virus more exposed to the response of the immune system and thus cytotoxic $CD8^+$ cells (cytotoxic T lymphocytes) or the antiviral agents administered show greater effectiveness. Considering this approach, Fonteh and Meyer[21] reported the *in vitro* reactivation of latent HIV-1 by using two bis(thiosemicarbazonate) gold(III) complexes (Figure 6.1, complexes **131** and **130**). The 50% cytotoxicity concentration (CC_{50}) values of the complexes were 0.53 ± 0.12 µM and 1.0 ± 0.4 µM for **131** and **130**, respectively, in U1 cells (HIV latency cell line). At non-toxic concentrations of 0.2 and 0.5 µM and in the absence of phorbol myristate acetate, complexes **131** and **130** reactivated the virus in U1 cells by 2.7- and 2.3-fold, respectively, and, in comparison, a 2.6-fold increase was observed in viral reactivation for hydroxyurea, which was used as a cytostatic and latent HIV reactivation control. In addition, the complexes did not induce viral reactivation during co-stimulation with phorbol myristate acetate, which indicated the absence of an additive effect. Complexes **131** and **130** also stimulated protein kinase C (PKC) activity by up to 32% with no noteworthy inhibition of histone deacetylase (HDAC). Overall, this research revealed the dual potential of these complexes in terms of being cytostatic and causing reactivation in the viral activation/elimination process.

N-Heterocyclic carbenes (NHCs) are constantly receiving attention for the development of metal complexes with exciting physico-chemical and biological properties. Metal complexes of NHCs with gold metal ions are quite fascinating owing to their efficacy as bioactive agents. The stability of the metal center with the NHC ligands is one of the main advantages of these complexes. This stability ensures slow release of the active therapeutic agent and, therefore, maintains the requisite concentration of the active agent.[22,23] Additionally, it is easy to functionalize NHCs through *N*-substituents, which allows tuning of their hydrophilic and lipophilic properties. Sanchez *et al.*[24] reported the *in vitro* anti-HIV activity of a bimetallic bridged gold complex (Figure 6.4, complex **138**) against HIV-1 in infected MT4 leukemia cells. The complex displayed substantial inhibition of the virus at low concentration.

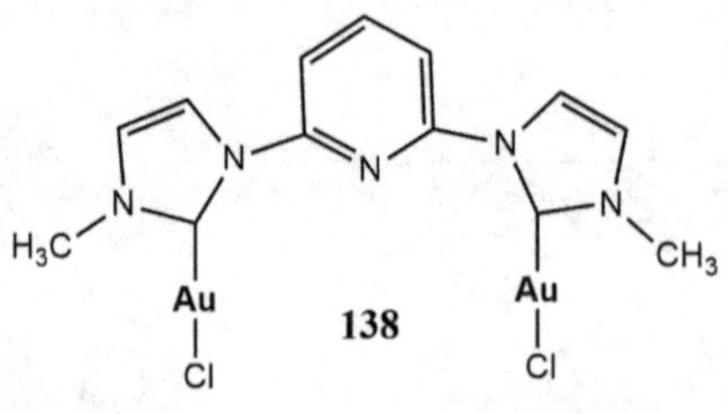

Figure 6.4 Chemical structure of bimetallic bridged gold complex (**138**).[24]

At 0.05, 0.5 and 1.0 μg mL^{-1}, the percentage inhibition of the virus-infected cells was equal to 81.9 ± 2.0, 73.8 ± 2.1 and 70.4 ± 4.6, respectively. The viral inhibitory activity of the complex was attributed to the interaction of the complex with the viral envelope proteins.

The recent pandemic of severe acute respiratory syndrome coronavirus 2 (SARS-CoV-2) led to a worldwide health emergency. More than 700 million people were affected globally due to this pandemic.[25,26] The whole world struggled to control the spread of this virus and it had a drastic impact on healthcare and the global economy. Vaccine and therapeutic development to combat this viral infection was undertaken worldwide. In this respect, Gil-Moles *et al.*[27] demonstrated the inhibition of the spike–angiotensin converting enzyme 2 (ACE2) interaction and papain-like protease (PLpro) activity of five gold complexes (Figure 6.5, complexes **139–143**). The complexes effectively inhibited the interaction of the SARS-CoV-2 spike protein with the ACE2 host receptor, and therefore interfered with the entry of the virus into the cell. In addition, the complexes efficiently inhibited PLpro of SARS-CoV-1 and SARS-CoV-2, an important enzyme involved in the replication of the virus, as well as removed zinc ions from the labile center of the PLpro enzyme. Overall, this research indicated the enormous potential of these gold complexes as anti-SARS-CoV-2 agents, and thus future investigations of these complexes as anti-SARS-CoV agents are warranted. Massai and co-workers[28] documented the inhibitory activities of five gold(I) complexes

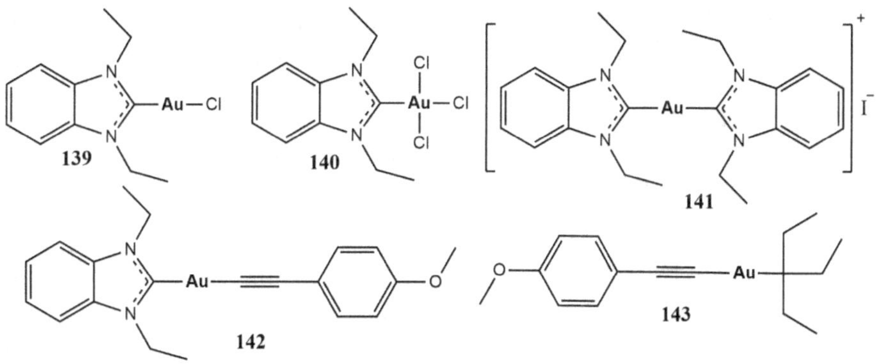

Figure 6.5 Chemical structures of gold complexes with NHC or alkynyl ligands (**139–143**).[27]

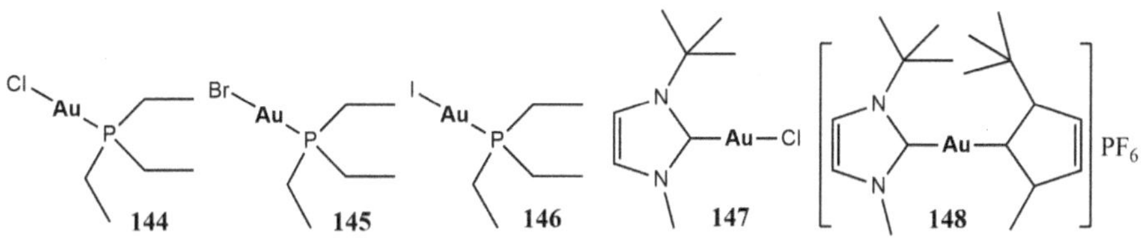

Figure 6.6 Chemical structures of gold complexes with halido and carbene ligands (**144–148**).[28]

(Figure 6.6, complexes **144–148**) against SARS-CoV-2 main protease (SARS-CoV-2 Mpro), which is an important biological target for the treatment of COVID-19 infection. Interestingly, all the complexes were potent inhibitors of SARS-CoV-2 Mpro, with inhibition constant (K_i) values in the range 2.1–0.4 μM. The complexes formed tight metal complex–protein adducts. The structures of the metal complex–Mpro adducts of complexes **144** and **145** revealed that gold coordination occurred at the level of the catalytic Cys 145 in complex **144**, and at both Cys 145 and Cys 156 in complex **145**. This coordination of gold atoms of the complexes to the functionally relevant cysteine residues was proposed as the molecular basis for the strong enzyme inhibitory properties of the complexes. Overall, this research suggests the importance of future investigation of the potential of these gold complexes for combating SARS-CoV-2 infection in *in vivo* models, for better understanding of their pharmacodynamics in relation to their inhibitory effects.

Chikungunya virus (CHIKV) is the causative agent of Chikungunya fever. CHIKV is transmitted and spread by the bite of female mosquitoes, *Aedes aegypti* and *Aedes albopictus*. In tropical and subtropical countries, Chikungunya fever is an endemic disease,[29,30] and the virus has also been identified in Mediterranean countries and North America.[31] Chikungunya fever is associated with symptoms of fever, fatigue, nausea, arthralgia and polyarthralgia. In addition, hepatitis, encephalopathy and myocarditis have been reported in some patients, which can lead to death.[32] In some patients, this viral infection progresses to a chronic condition, which results in disability.[33] Unfortunately, none of the antiviral drugs inhibit CHIKV, and this causes huge health and economic impacts. In this context, Aires *et al.*[34] reported the anti-CHIKV effects of four gold(I) complexes of triphenylphosphine ligands (Figure 6.7, complexes **149–152**). The complexes displayed viral inhibition in the range 50–99%. The complexes showed binding to *N*-acetyl-L-cysteine. The most remarkable inhibition was reported in post-entry activity, wherein the complexes blocked the replication of the virus by up to 99%, which was followed by inhibition of up to 95% of the early stages of infection. The complexes significantly interacted with double-stranded (ds) RNA, which is an important biological target for inhibition of CHIKV. Considering all these findings, complex **149** was identified as a promising lead for the future development of anti-CHIKV agents.

Human adenovirus (HAdV) is known to cause serious diseases in recipients of hematopoietic stem cell transplants, particularly in pediatric patients. HAdV has been identified as the main causative agent of community-acquired pneumonia.[35] At present, there are no effective therapeutic medicines for the treatment of HAdV infections, and, therefore, the identification and development of safe and effective anti-HAdV agents is needed.[36,37] Although cidofovir is frequently prescribed for treating HAdV infections,[38] its medicinal use suffers from drawbacks such as low efficacy, less bioavailability and several adverse side effects.[39] In addition, the use of brincidofovir, which is a lipidic conjugate of cidofovir, is also limited owing to its similar long-term adverse effects to cidofovir.[40,41] In response to this, Balsera-Manzanero and

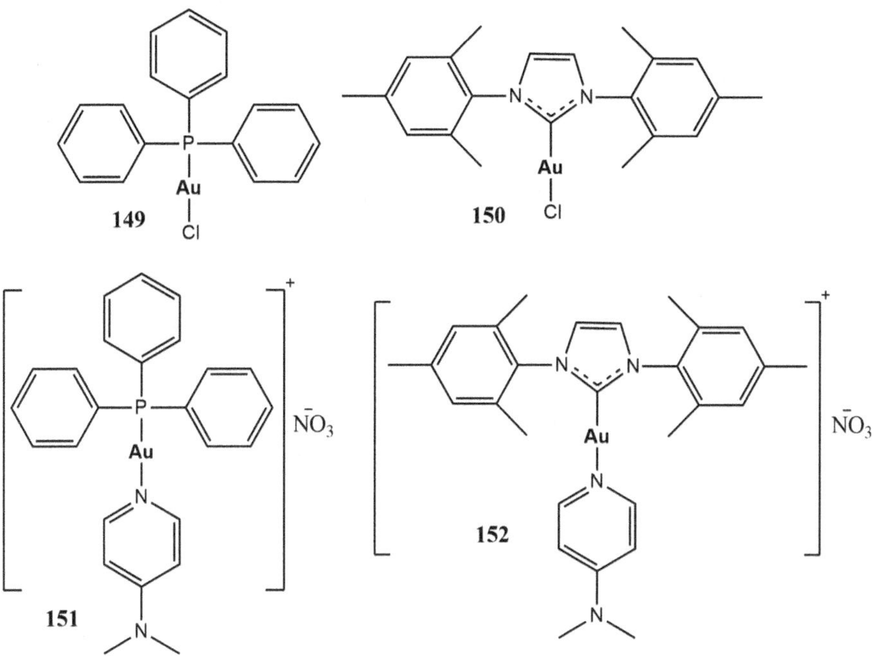

149 **150** **151** **152**

Figure 6.7 Chemical structures of gold(I) complexes of triphenylphosphine ligands (**149–152**).[34]

co-workers[42] reported the anti-HAdV properties of submicromolar concentrations of four heteroleptic (C^S)-cycloaurated complexes (Figure 6.8, complexes **153–156**). The complexes displayed low cytotoxicity and possessed anti-HAdV activity. The antiviral activity of the complexes was attributed to their targeting of the stages of viral replication cycle following DNA replication. Interestingly, the complexes exhibited strong inhibition of DNA replication of human cytomegalovirus (HCMV). Complex **155** had the best selectivity index (22.76) and the highest anti-HCMV activity (90% inhibition). Complex **155** could be used as a guiding template for the design and development of antiviral gold complexes until a suitable complex enters into the preclinical setting.

An overall assessment of the research reviewed in this chapter reveals that several gold(I) and gold(III) complexes with a diverse range of ligand systems have been reported with interesting antiviral properties. Bis(thiosemicarbazonate) ligands, 1,10-bis(diphenylphosphino)metallocene derivatives, NHCs, triphenylphosphines, heteroleptic (S^C)-cyclometalated ligands and other ligand systems have been employed for the rational design and development of gold complexes with antiviral properties. Some of the reported complexes have shown promising antiviral properties against several disease-causing viruses, such as HIV, coronavirus, Chikungunya virus and HCMV. Importantly, some complexes have shown much better *in vitro*

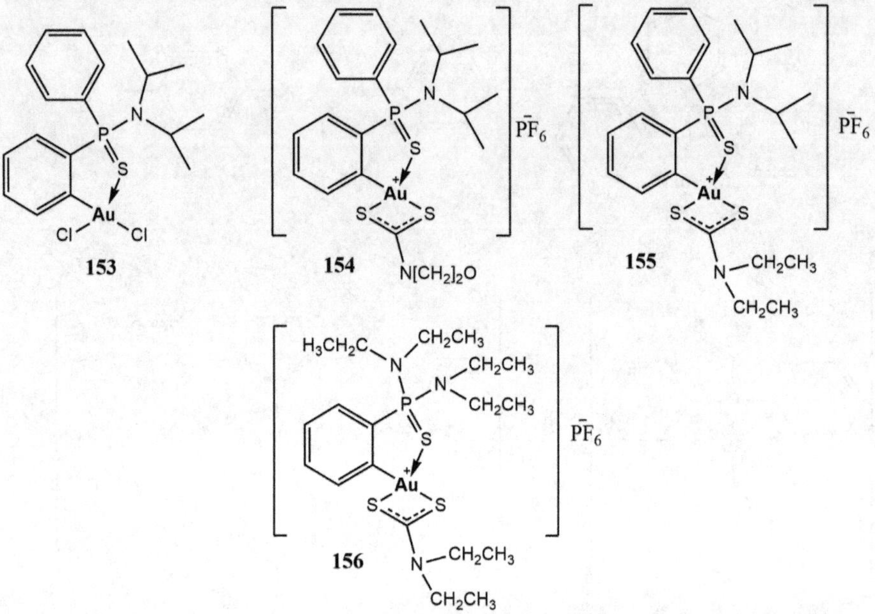

Figure 6.8 Chemical structures of heteroleptic (C^S)-cycloaurated complexes (**153–156**).[42]

antiviral properties in comparison to the standard antiviral drug azidothymidine. In addition, some complexes have shown glimpses of behavior as virostatic agents. Additionally, several complexes have shown promise for being cytostatic and causing reactivation in the viral activation/elimination process. Overall, the antiviral gold complexes described in this chapter have shown the huge potential of these complexes for the design and development of future antiviral agents.

References

1. J. E. Pope, P. Hong and B. E. Koehler, *J. Rheumatol.*, 2002, **29**, 255–260.
2. R. E. de Paiva, A. M. Neto, I. A. Santos, A. C. Jardim, P. P. Corbi and F. R. Bergamini, *Dalton Trans.*, 2020, **49**, 16004–16033.
3. K. Yamaguchi, H. Ushijima, M. Hisano, Y. Inoue, T. Shimamura, T. Hirano and W. E. G. Müller, *Microbiol. Immunol.*, 2001, **45**, 549–555.
4. M. G. Lewis, S. DaFonseca, N. Chomont, A. T. Palamara, M. Tardugno, A. Mai, M. Collins, W. L. Wagner, J. Yalley-Ogunro and J. Greenhouse, *AIDS*, 2011, **25**, 1347–1356.
5. P. Fonteh and D. Meyer, *Metallomics*, 2009, **1**, 427–433.
6. A. Marques, S. A. Carabineiro, M. Aureliano and L. Faleiro, *Toxics*, 2023, **11**, 879.
7. S. Payagala and A. Pozniak, *Clin. Dermatol.*, 2024, **42**, 119–127.

8. T. Okada, B. K. Patterson, S.-Q. Ye and M. E. Gurney, *Virology*, 1993, **192**, 631–642.
9. P. N. Fonteh, F. K. Keter and D. Meyer, *J. Inorg. Biochem.*, 2011, **105**, 1173–1180.
10. H. Bjelosevic, I. A. Guzei, L. C. Spencer, T. Persson, F. H. Kriel, R. Hewer, M. J. Nell, J. Gut, C. E. Van Rensburg and P. J. Rosenthal, *J. Organomet. Chem.*, 2012, **720**, 52–59.
11. N. Gama, K. Kumar, E. Ekengard, M. Haukka, J. Darkwa, E. Nordlander and D. Meyer, *BioMetals*, 2016, **29**, 389–397.
12. R. F. Siliciano and W. C. Greene, *Cold Spring Harbor Perspect. Med.*, 2011, **1**, a007096.
13. T.-W. Chun, L. Stuyver, S. B. Mizell, L. A. Ehler, J. A. M. Mican, M. Baseler, A. L. Lloyd, M. A. Nowak and A. S. Fauci, *Proc. Natl. Acad. Sci. U. S. A.*, 1997, **94**, 13193–13197.
14. D. Finzi, M. Hermankova, T. Pierson, L. M. Carruth, C. Buck, R. E. Chaisson, T. C. Quinn, K. Chadwick, J. Margolick, R. Brookmeyer, J. Gallant, M. Markowitz, D. D. Ho, D. D. Richman and R. F. Siliciano, *Science*, 1997, **278**, 1295–1300.
15. J. K. Wong, M. Hezareh, H. F. Günthard, D. V. Havlir, C. C. Ignacio, C. A. Spina and D. D. Richman, *Science*, 1997, **278**, 1291–1295.
16. T.-W. Chun, D. Finzi, J. Margolick, K. Chadwick, D. Schwartz and R. F. Siliciano, *Nat. Med.*, 1995, **1**, 1284–1290.
17. T. Pierson, J. McArthur and R. F. Siliciano, *Annu. Rev. Immunol.*, 2000, **18**, 665–708.
18. V. Quivy, E. Adam, Y. Collette, D. Demonte, A. Chariot, C. Vanhulle, B. Berkhout, R. Castellano, Y. De Launoit, A. Burny, J. Piette, V. Bours and C. Van Lint, *J. Virol.*, 2002, **76**, 11091–11103.
19. L. Colin and C. Van Lint, *Retrovirology*, 2009, **6**, 111.
20. W. Maek-A-Nantawat, S. Buranapraditkun, J. Klaewsongkram and K. Ruxrungthum, *Viral Immunol.*, 2007, **20**, 66–75.
21. P. Fonteh and D. Meyer, *BMC Infect. Dis.*, 2014, **14**, 680.
22. W. J. Youngs, A. R. Knapp, P. O. Wagers and C. A. Tessier, *Dalton Trans.*, 2012, **41**, 327–336.
23. T. J. Siciliano, M. C. Deblock, K. M. Hindi, S. Durmus, M. J. Panzner, C. A. Tessier and W. J. Youngs, *J. Organomet. Chem.*, 2011, **696**, 1066–1071.
24. O. Sánchez, S. González, Á. R. Higuera-Padilla, Y. León, D. Coll, M. Fernández, P. Taylor, I. Urdanibia, H. R. Rangel and J. T. Ortega, *Polyhedron*, 2016, **110**, 14–23.
25. P. Zhou, X.-L. Yang, X.-G. Wang, B. Hu, L. Zhang, W. Zhang, H.-R. Si, Y. Zhu, B. Li and C.-L. Huang, *Nature*, 2020, **579**, 270–273.
26. C. Wang, P. W. Horby, F. G. Hayden and G. F. Gao, *Lancet*, 2020, **395**, 470–473.
27. M. Gil-Moles, U. Basu, R. Büssing, H. Hoffmeister, S. Türck, A. Varchmin and I. Ott, *Chem. – Eur. J.*, 2020, **26**, 15140–15144.
28. L. Massai, D. Grifagni, A. De Santis, A. Geri, F. Cantini, V. Calderone, L. Banci and L. Messori, *Biomolecules*, 2022, **12**, 1675.

29. C. Ngoagouni, B. Kamgang, M. Kazanji, C. Paupy and E. Nakouné, *Parasites Vectors*, 2017, **10**, 164.
30. M. Mavale, D. Parashar, A. Sudeep, M. Gokhale, Y. Ghodke, G. Geevarghese, V. Arankalle and A. C. Mishra, *Am. J. Trop. Med. Hyg.*, 2010, **83**, 1242.
31. L. A. Silva and T. S. Dermody, *J. Clin. Invest.*, 2017, **127**, 737–749.
32. T. Das, M. C. Jaffar-Bandjee, J. J. Hoarau, P. K. Trotot, M. Denizot, G. Lee-Pat-Yuen, R. Sahoo, P. Guiraud, D. Ramful and S. Robin, *Prog. Neurobiol.*, 2010, **91**, 121–129.
33. L. Dupuis-Maguiraga, M. Noret, S. Brun, R. Le Grand, G. Gras and P. Roques, *PLoS Neglected Trop. Dis.*, 2012, **6**, e1446.
34. R. L. Aires, I. A. Santos, J. V. Fontes, F. R. Bergamini, A. C. G. Jardim and C. Abbehausen, *Metallomics*, 2022, **14**, mfac056.
35. K. M. MacNeil, M. J. Dodge, A. M. Evans, T. M. Tessier, J. B. Weinberg and J. S. Mymryk, *Trends Mol. Med.*, 2023, **29**, 4–19.
36. M. J. Dodge, K. M. MacNeil, T. M. Tessier, J. B. Weinberg and J. S. Mymryk, *Antiviral Res.*, 2021, **188**, 105034.
37. B. Saha and R. J. Parks, *Microorganisms*, 2020, **8**, 1284.
38. J. P. S. Begum, L. Ngangom, P. Semwal, S. Painuli, R. Sharma and A. Gupta, *Hum. Cell*, 2023, **36**, 877–893.
39. M. Moshirfar, N. Kelkar, T. Peterson, J. Bradshaw, L. Parker, Y. C. Ronquillo and P. C. Hoopes, *Eye & Contact Lens*, 2023, **49**, 127–134.
40. M. S. Grimley, G. M. Marón, V. K. Prasad, D. A. Jacobsohn, J.-A. H. Young, G. Chittick, T. M. Brundage and H. Momméja-Marin, *Biol. Blood Marrow Transplant.*, 2015, **21**, S108–S109.
41. G. Chittick, M. Morrison, T. Brundage and W. G. Nichols, *Antiviral Res.*, 2017, **143**, 269–277.
42. M. Balsera-Manzanero, R. G. Soengas, M. Carretero-Ledesma, C. Ratia, M. J. Iglesias, J. Pachón, F. López-Ortiz, E. Cordero, S. M. Soto and J. Sánchez-Céspedes, *Heliyon*, 2024, **10**, e2701.

CHAPTER 7

Gold Complexes as Antiparasitic Agents

Malaria has been a challenging disease for humans for a very long time. It has led to huge economic losses the world over. Malaria is a huge problem both clinically and economically since it is more prevalent in poorer countries of the world and so further hinders the social and economic development of those countries. The protozoan parasites of the genus *Plasmodium* are the causative agents of malaria. *Plasmodium falciparum, Plasmodium vivax, Plasmodium malariae, Plasmodium ovale* and *Plasmodium knowlesi* are the species that are known to infect humans and cause malaria. For effective treatment of malaria, several therapeutic antimalarial drugs have been discovered and developed over time. Recent updates in the literature of bioactive gold complexes reveal that several gold complexes with exciting antiparasitic properties have been reported.[1–3] Wash and co-workers[4] reported the antiplasmodial activity of gold(III) complexes of the antimalarial drugs amodiaquine and primaquine against *P. falciparum*. The complexes were as active as the standard drugs. This finding provided the necessary impetus for the rational design and development of antimalarial gold complexes. Fontinha *et al.*[5] documented the antiplasmodial properties of six gold(III) bis(dithiolene) complexes (Figure 7.1, complexes **157–162**). The complexes exhibited exciting antiplasmodial activity against the liver stages of *Plasmodium berghei*. The antiplasmodial activity of the complexes was higher than that of primaquine. Moreover, complex **157** was quite stable under biological conditions, which further added to its therapeutic profile as a drug.

Trypanosomes belonging to *Trypanosoma brucei* species are parasitic eukaryotes that cause severe illnesses in both humans and animals.[6–9] In humans, *T. brucei* parasites cause human African trypanosomiasis (HAT). HAT (or sleeping sickness) is a lethal disease for humans if left untreated.[10,11]

The Therapeutic Potential of Gold Complexes
By Waseem A. Wani, Mohamed F. Alajmi, Afzal Hussain and Irshad Ahmad Wani

Published by the Royal Society of Chemistry, www.rsc.org

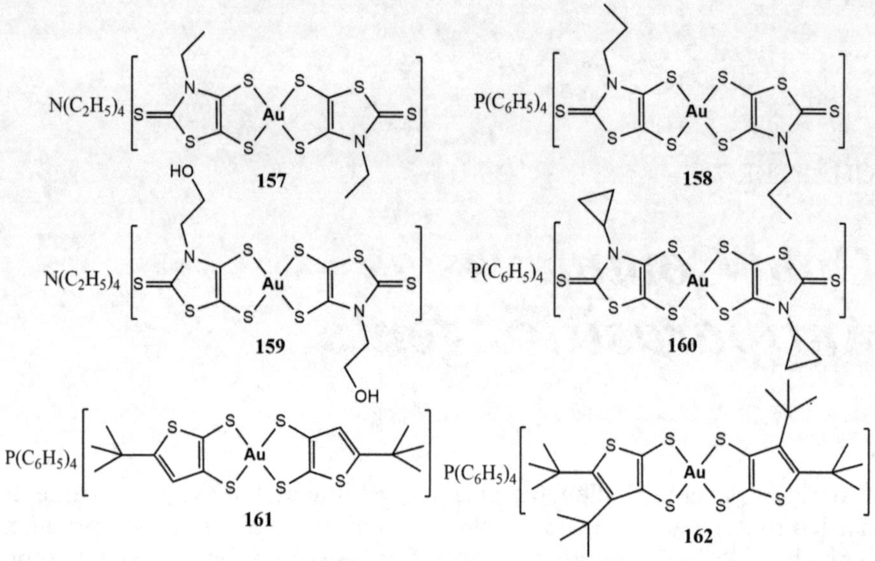

Figure 7.1 Chemical structures of gold(III)-bis(dithiolene) complexes (**157–162**).[5]

At present, only a few drugs are available for treating HAT[12] and the available drugs have high costs and are generally associated with deleterious side effects.[13,14] In an effort to rationally develop agents for the treatment of the parasitic infections due to *T. brucei*, Winter *et al.*[15] reported the anti-trypanosomal activity of two gold(I)-N-heterocyclic carbene (NHC) complexes (Figure 7.2, complexes **163** and **164**) against *T. brucei*. The complexes showed promising anti-trypanosomal activity with IC_{50} values in the range 0.9–3.0 nM. In addition, the complexes enabled the detachment of flagella from the parasitic cells. The complexes also induced complete degeneration of the cytoskeleton of the parasites, wherein only the microtubules of the detached flagella stayed intact. Interestingly, the complexes displayed selectivity towards *T. brucei* compared to human HeLa cells. Overall, these complexes may be envisaged as potential anti-trypanosomal agents.

Chagas disease, caused by *Trypanosoma cruzi*, is a grave public health issue owing to inadequate treatment results due to drug resistance to the available drugs and their deleterious side effects. This disease remains a neglected disease and is endemic in Latin America. In order to develop effective therapeutic agents for the treatment of Chagas disease, Lopes *et al.*[16] demonstrated the *in vitro* and *in vivo* antiparasitic activity of an organometallic gold(III) complex bearing a S,N,S-donating thiosemicarbazone (Figure 7.3, complex **165**). The complex was more active than benznidazole in causing the elimination of extracellular trypomastigotes and intracellular amastigotes of *T. cruzi*, without any associated cytotoxic effects towards mammalian cells. In *in vivo* investigations, the complex in phosphate-buffered saline solution became neutral and reduced parasitemia, along with tissue parasitism. In addition,

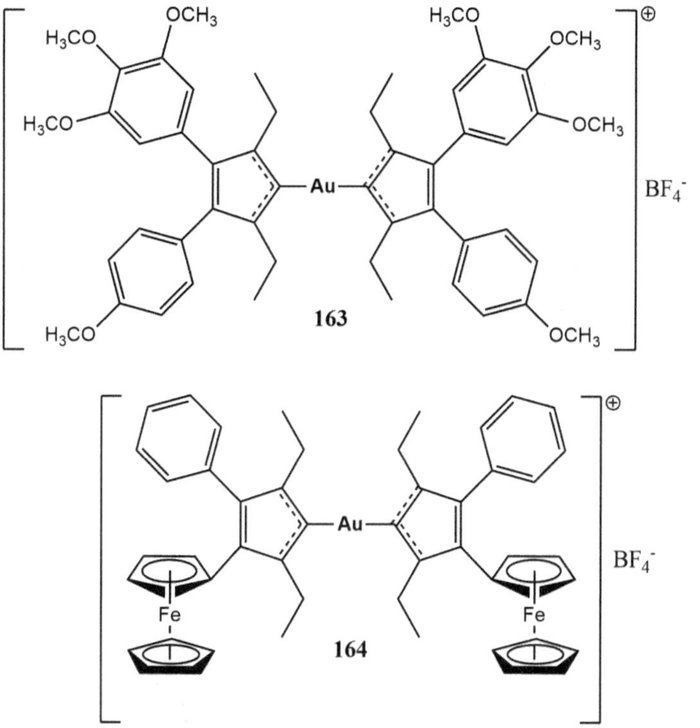

Figure 7.2 Chemical structures of gold(I)-NHC complexes (**163** and **164**).[15]

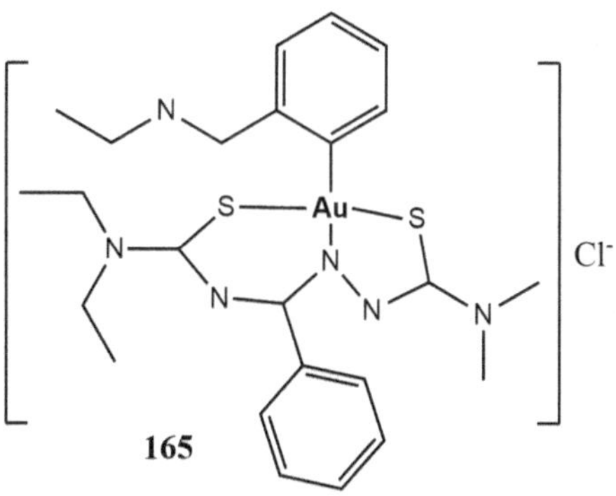

Figure 7.3 Chemical structures of an organometallic gold(III) complex bearing a S,N,S-donating thiosemicarbazone (**165**).[16]

the complex did not harm the liver and heart of experimental mice, through tissue damage, at a dose of 2.8 mg kg^{-1} day^{-1}. A 100% survival of the mice was reported during treatment with the complex during the acute phase. The parasite load in the surviving mice after 150 days of infection was equal to the parasitic dose in the mice treated with benznidazole. Interestingly, modulation of interferon-gamma levels was identified, which might have targeted positive outcomes of the disease. This research clearly indicated the huge promise of the reported complex in an *in vivo* experimental mouse model against Chagas disease. Salsi and co-workers[17] documented the antiparasitic activity of organometallic gold(III) complexes containing tridentate halogen-substituted thiosemicarbazones (Figure 7.4, complexes **166–173**) against *T. cruzi*. The complexes displayed high antiparasitic activity at low micromolar concentrations, and were statistically at par with benznidazole in terms of activity. However, complex **167**, containing a chlorine substituent, showed mild activity. Ordenes-Rojas and co-workers[18] reported the *in vitro* antiparasitic activities of four gold(I) complexes containing bioactive 5-nitrofuryl-containing thiosemicarbazones (Figure 7.5, complexes **174–177**) against the trypomastigote form of *T. cruzi* and selectivity towards human

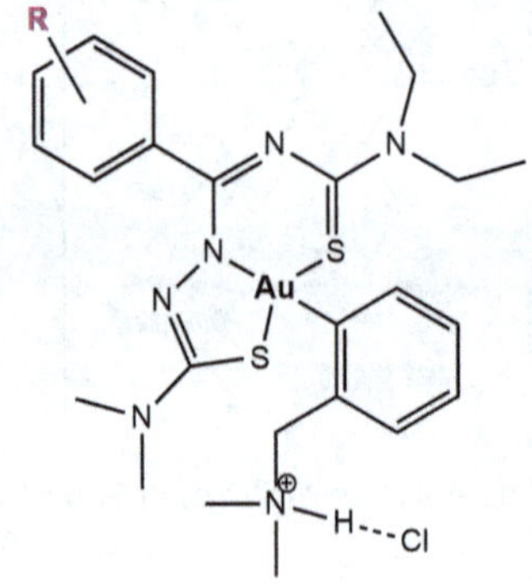

Complex 166: R = 4-H
Complex 167: R = 4-Cl
Complex 168: R = 4-Br
Complex 169: R = 4-F
Complex 170: R = 3-F
Complex 171: R = 4-CF_3
Complex 172: R = 3-CF_3
Complex 173: R = 3,5-diF

Figure 7.4 Chemical structures of organometallic gold(III) complexes containing tridentate halogen-substituted thiosemicarbazones (**166–173**).[17]

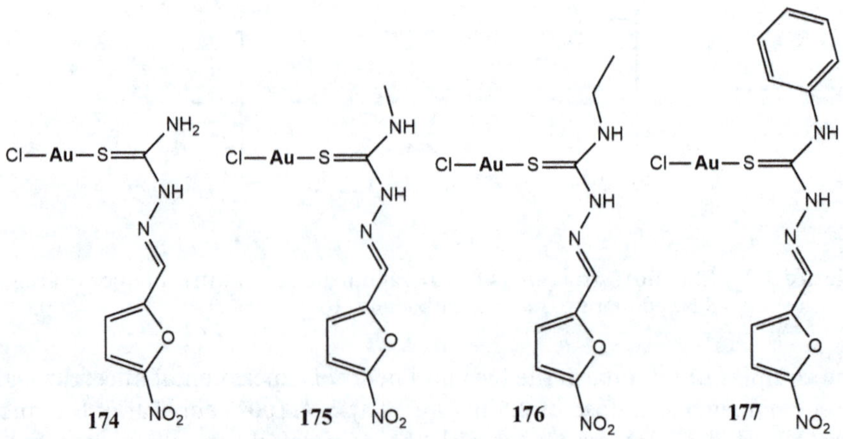

174 **175** **176** **177**

Figure 7.5 Chemical structures of gold(I) complexes containing bioactive 5-nitrofuryl-containing thiosemicarbazones (**174–177**).[18]

cells of endothelial morphology. The complexes were active against *T. cruzi*, with IC_{50} values in the micromolar range. Complexes **175** and **176** were the most active complexes, with activities comparable to that of the standard drug nifurtimox. However, the two complexes were moderately selective towards normal cells. All the complexes elevated intracellular reactive oxygen species (ROS). The maximum content of intracellular ROS was generated by complex **176**, which correlated with it having the lowest IC_{50} value among the tested compounds. The overall findings indicated that the ROS-generating ability of the complexes was the mechanism for their antiparasitic activity.

The causative agents for leishmaniasis are *Leishmania* protozoan parasites. These parasites cause cutaneous or visceral manifestations. *Leishmania amazonensis* and *Leishmania braziliensis* are the key species of the parasite that are responsible for cutaneous leishmaniasis in Brazil.[19] The available drugs for the treatment of leishmaniasis have high costs, low efficacy and toxicity issues, which demands the design and development of new anti-leishmanial agents. In a quest to develop agents for the treatment of leishmaniasis, Chaves and co-workers[20] reported the anti-leishmanial properties of gold(I) complexes with 5-phenyl-1,3,4-oxadiazole-2-thione and phosphine (Figure 7.6, complexes **178–187**). The complexes were active against *Leishmania infantum* intracellular amastigotes, with IC_{50} values in the range 0.9–4.2 μM. The complex **183** was highly selective against *L. infantum* intracellular amastigotes along with promising activity towards the parasite. Interestingly, the triethylphosphine-gold(I) complexes (**183–187**) presented better anti-leishmanial activity than the triphenylphosphine complexes (**178–182**), in addition to displaying antimony cross-resistance in *Leishmania guyanensis*. These findings revealed the potential of these complexes as important chemical tools for better understanding the action mechanisms and resistance patterns for further

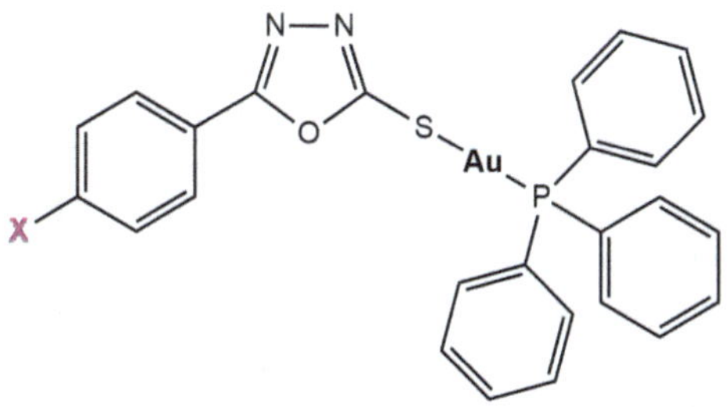

Complex 178: X = H
Complex 179: X = Cl
Complex 180: X = F
Complex 181: X = NO_2
Complex 182: X = OCH_3

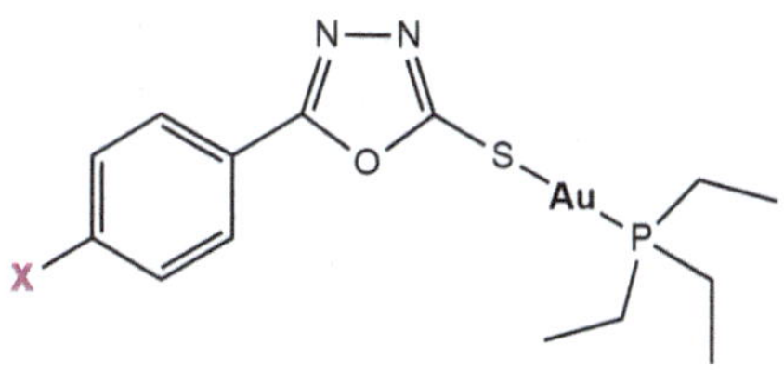

Complex 183: X = H
Complex 184: X = Cl
Complex 185: X = F
Complex 186: X = NO_2
Complex 187: X = OCH_3

Figure 7.6 Chemical structures of gold(I) complexes with 5-phenyl-1,3,4-oxadiazole-2-thione and phosphine (**178–187**).[20]

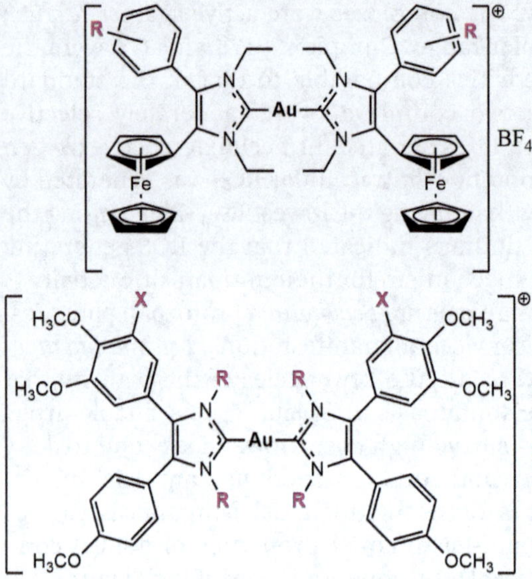

Complex 188: R = H
Complex 189: R = 3,4,5-OCH_3
Complex 190: R = 4-OCH_3

Complex 191: R = CH_3, X = OCH_3
Complex 192: R = CH_2CH_3, X = OCH_3
Complex 193: R = CH_3, X = Cl
Complex 194: R = CH_2CH_3, X = Br

Figure 7.7 Chemical structures of cationic gold(I)-carbene complexes with various 4,5-diarylimidazolylidene ligands (**188–194**).[21]

drug development. Koko *et al.*[21] demonstrated the antiparasitic properties of cationic gold(I)-carbene complexes with various 4,5-diarylimidazolylidene ligands (Figure 7.7, complexes **188–194**) against *Leishmania major*. Complex **188** displayed the best activities towards *L. major* amastigotes. In addition, the complex showed distinct selectivity for *Trypanosoma gondii* cells. Interestingly, the ferrocene complexes **188–190** displayed more activity against the *L. major* amastigotes and the *T. gondii* tachyzoites, in comparison to the other anisyl-gold complexes tested (**191–194**) and the standard drugs, atovaquone and amphotericin B. Moreover, complexes **188** and **192** had the highest selectivities for *L. major* amastigotes, whereas the complexes **191** and **193** displayed the highest selectivities for *L. major* promastigotes, with **193** being the most active complex against *L. major* promastigotes. Minori and co-workers[22] reported the *in vitro* anti-leishmanial activity of a series of mononuclear organometallic Au(I)/Au(III) complexes (Figure 7.8, complexes **195–203**) against promastigotes and amastigotes of *L. amazonensis* and *L. braziliensis*. The complex **197** demonstrated a low EC_{50} value (4.24 ± 0.23 μM) in promastigote cells, with no toxicity in host macrophages. Additionally, complexes **200** and **201** demonstrated strong activities against intracellular amastigotes from *L. amazonensis*. The mechanistic studies of the complexes indicated that the complexes affected the plasma membrane integrity of *L. amazonensis* cells. Tunes *et al.*[23] documented exciting anti-leishmanial activity of gold(I)-derived complexes (Figure 7.9, complexes **204–207**) against *L. infantum* and *L. braziliensis* intracellular amastigotes, with IC_{50} values ranging from 0.5 to 5.5 μM. In addition, the gold(I) complexes potently inhibited trypanothione reductase enzyme, with IC_{50} values ranging from 1 to 7.8 μM. Triethylphosphine-based

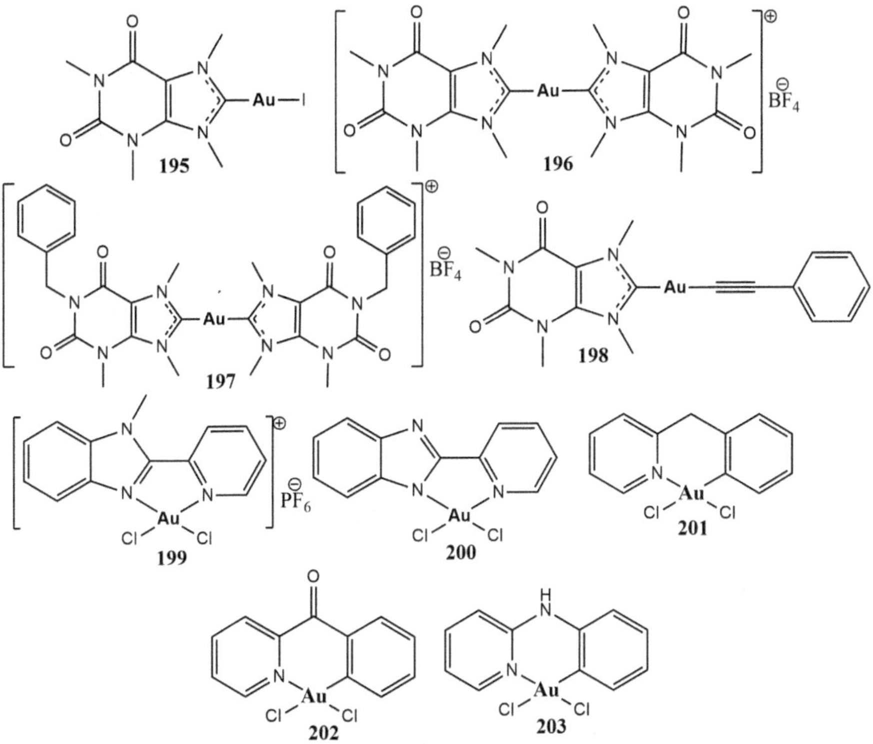

Figure 7.8 Chemical structures of mononuclear organometallic Au(I)/Au(III) complexes (**195–203**).[22]

complexes elevated ROS production and reduced mitochondrial respiration after exposure for 2 h. This indicated that the reported gold(I) complexes led to oxidative stress by direct production of ROS. Interestingly, no cross-resistance to antimony was observed, and the antimony-resistant mutants were more sensitive to some complexes. During *in vivo* assays, reduced lesion size and parasite burden were reported in BALB/c mice infected with luciferase-expressing *L. braziliensis* or *L. amazonensis*, upon treatment with orally administered 12.5 mg kg^{-1} day^{-1} of complex **206** or complex **207**. In addition, the synergistic combination of complex **206** with miltefosine decreased miltefosine treatment time by 50%. Overall, the complexes **206** and **207** displayed promising pharmacokinetic and toxicity profiles, which encourages the further investigation of these complexes. Moreover, these complexes show huge promise as anti-leishmanial agents, particularly for the treatment of leishmaniasis due to antimony-resistant parasites.

A critical analysis of the research discussed here highlights that a number of gold(I) and gold(III) complexes with an array of differently structured ligands have been reported with encouraging antiparasitic properties. NHCs, phosphine ligands, thiosemicarbazones, bis(dithiolene) ligands, phosphole ligands and other ligand systems have all been utilized for

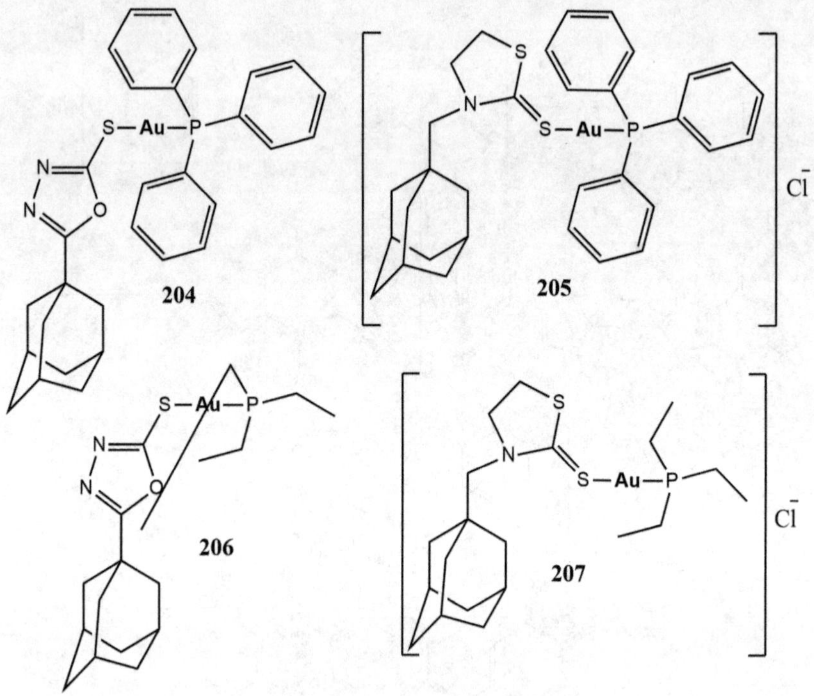

Figure 7.9 Chemical structures of gold(I)-derived complexes (**204–207**).[23]

the rational design and development of gold complexes with antiparasitic properties. Some of the complexes described in this chapter have exhibited promising antiparasitic properties against several pathogenic parasites, including *P. berghei*, *T. brucei*, *T. cruzi*, *L. infantum*, *L. major*, *L. amazonensis*, *L. braziliensis*, *Leishmania donovani* and *T. gondii*. Interestingly, some complexes displayed better antiparasitic properties than the standard market-available antiparasitic drugs, such as primaquine, benznidazole, nifurtimox, atovaquone and amphotericin B. In addition, several complexes displayed antimony cross-resistance in certain species of *Leishmania*, and such complexes might serve as important chemical tools for gaining insight into the mechanisms of action and resistance patterns, for further drug development. Overall, the antiparasitic gold complexes discussed in this chapter have displayed huge potential for the design and development of future antiparasitic agents.

References

1. M. Navarro, *Coord. Chem. Rev.*, 2009, **253**, 1619–1626.
2. I. S. Al Nasr, M. Weise, W. S. Koko, T. A. Khan, R. Schobert and B. Biersack, *Appl. Organomet. Chem.*, 2023, **37**, e6952.

3. L. Feng, S. Pomel, P. Latre de Late, A. Taravaud, P. M. Loiseau, L. Maes, F. Cho-Ngwa, C. A. Bulman, C. Fischer and J. A. Sakanari, *Molecules*, 2020, **25**, 5075.
4. N. Wasi, H. B. Singh, A. Gajanana and A. N. Raichowdhary, *Inorg. Chim. Acta*, 1987, **135**, 133–137.
5. D. Fontinha, S. A. Sousa, T. S. Morais, M. Prudêncio, J. H. Leitão, Y. Le Gal, D. Lorcy, R. A. Silva, M. F. Velho and D. Belo, *Metallomics*, 2020, **12**, 974–987.
6. R. Lucius and B. Loos-Frank, *Parasitologie: Grundlagen für Biologen, Mediziner und Veterinärmediziner*, Spektrum Akademischer Verlag, Heidelberg, 1997.
7. D. H. Smith, J. Pepin and A. H. Stich, *Br. Med. Bull.*, 1998, **54**, 341–355.
8. J. Pépin and H. A. Méda, *Adv. Parasitol.*, 2001, **49**, 71–132.
9. D. A. Maslov, S. A. Podlipaev and J. Lukes, *Mem. Inst. Oswaldo Cruz*, 2001, **96**, 397–402.
10. M. P. Barrett and A. H. Fairlamb, *Parasitol. Today*, 1999, **15**, 136–140.
11. M. P. Barrett, *Lancet*, 1999, **353**, 1113–1114.
12. D. Steverding, *Parasites Vectors*, 2008, **1**, 3.
13. M. P. Barrett, D. W. Boykin, R. Brun and R. R. Tidwell, *Br. J. Pharmacol.*, 2007, **152**, 1155–1171.
14. T. N. Baral, *J. Biomed. Biotechnol.*, 2010, **2010**, 1–24.
15. I. Winter, J. Lockhauserbäumer, G. Lallinger-Kube, R. Schobert, K. Ersfeld and B. Biersack, *Mol. Biochem. Parasitol.*, 2017, **214**, 112–120.
16. C. D. Lopes, B. Possato, A. P. S. Gaspari, R. J. Oliveira, U. Abram, J. P. A. Almeida, F. D. R. Rocho, A. Leitão, C. A. Montanari, P. I. S. Maia, J. S. Da Silva, S. De Albuquerque and Z. A. Carneiro, *ACS Infect. Dis.*, 2019, **5**, 1698–1707.
17. F. Salsi, G. Bulhões Portapilla, K. Schutjajew, M. Roca Jungfer, A. Goulart, A. Hagenbach, S. De Albuquerque and U. Abram, *Eur. J. Inorg. Chem.*, 2019, **2019**, 4455–4462.
18. J. Órdenes-Rojas, P. Risco, J. Ortega-Campos, G. Barriga-González, A. Liempi, U. Kemmerling, D. Gambino, L. Otero, C. Olea Azar and E. Rodríguez-Arce, *Inorganics*, 2024, **12**, 133.
19. J. Alvar, I. D. Vélez, C. Bern, M. Herrero, P. Desjeux, J. Cano, J. Jannin, M. den Boer and W. L. C. Team, *PLoS One*, 2012, **7**, e35671.
20. J. D. S. Chaves, L. G. Tunes, C. H. de J. Franco, T. M. Francisco, C. C. Correa, S. M. Murta, R. L. Monte-Neto, H. Silva, A. P. S. Fontes and M. V. de Almeida, *Eur. J. Med. Chem.*, 2017, **127**, 727–739.
21. W. S. Koko, J. Jentzsch, H. Kalie, R. Schobert, K. Ersfeld, I. S. Al Nasr, T. A. Khan and B. Biersack, *Arch. Pharm.*, 2020, **353**, e1900363.
22. K. Minori, L. B. Rosa, R. Bonsignore, A. Casini and D. C. Miguel, *ChemMedChem*, 2020, **15**, 2146–2150.
23. L. G. Tunes, R. E. Morato, A. Garcia, V. Schmitz, M. Steindel, J. D. Corrêa-Junior, H. F. Dos Santos, F. Frézard, M. V. De Almeida, H. Silva, N. S. Moretti, A. L. B. De Barros and R. L. Do Monte-Neto, *ACS Infect. Dis.*, 2020, **6**, 1121–1139.

CHAPTER 8

Gold Complexes as Anti-inflammatory Agents

The clinical use of gold(I) complexes, such as auranofin, as anti-inflammatory agents[1–3] served as a great stimulus for the rational design and development of other anti-inflammatory gold complexes. Current investigations of the medicinal potential of gold complexes can be partly attributed to the clinical success of anti-inflammatory gold(I) complexes, and partly to the inimitable chemistry of gold in the physiological environment, particularly the high affinity of gold for thiols and selenols of proteins, which are important components of the biological targets of various diseases, including inflammatory diseases.[4] The literature describes investigations of gold(I) and gold(III) complexes for varied ligand systems, with interesting anti-inflammatory properties.[2,5,6] Trávníček *et al.*[2] reported the *in vitro* and *in vivo* anti-inflammatory activity of gold(I) complexes based on six-substituted purine ligands (Figure 8.1, complexes **208–215**) against lipopolysaccharide (LPS)-activated macrophages derived from the THP-1 cell line and carrageenan-induced hind paw edema model in rats, respectively. The complexes **208, 210** and **213** strongly reduced the generation of pro-inflammatory cytokines, namely, tumor necrosis factor alpha (TNF-α), interleukin-1 beta (IL-1β) and high mobility group box 1 (HMGB1), without influencing production of the anti-inflammatory cytokine, interleukin-1 receptor antagonist (IL-1RA) in the LPS-activated macrophages. In addition, the complexes remarkably affected the formation of edema that was triggered by the intraplantar administration of polysaccharide λ-carrageenan to rats *in vivo*. Interestingly, all the complexes exhibited similar or pronounced biological properties in comparison to auranofin. The overall findings reported in this research indicated that the complexes were very effective anti-inflammatory agents and, thus, might be useful for treating inflammatory conditions such

The Therapeutic Potential of Gold Complexes
By Waseem A. Wani, Mohamed F. Alajmi, Afzal Hussain and Irshad Ahmad Wani

Published by the Royal Society of Chemistry, www.rsc.org

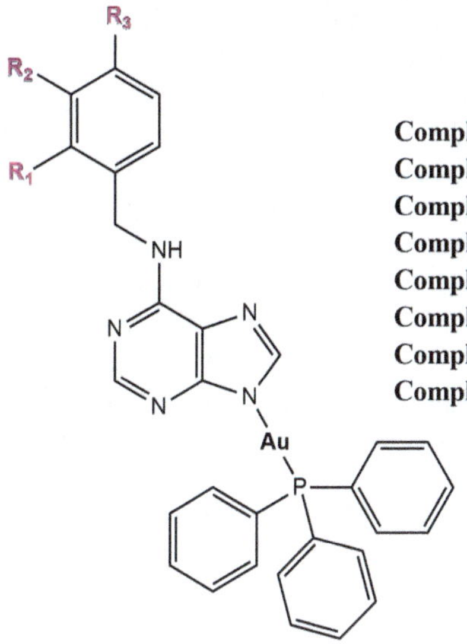

Complex 208: R_1 = F, R_2 = H and R_3 = H
Complex 209: R_1 = H, R_2 = F and R_3 = H
Complex 210: R_1 = Cl, R_2 = H and R_3 = H
Complex 211: R_1 = H, R_2 = Cl and R_3 = H
Complex 212: R_1 = H, R_2 = H and R_3 = Cl
Complex 213: R_1 = H, R_2 = CH_3 and R_3 = H
Complex 214: R_1 = F, R_2 = H and R_3 = CH_3
Complex 215: R_1 = F, R_2 = OCH_3 and R_3 = H

Figure 8.1 Chemical structures of gold(I) complexes based on six-substituted purine ligands (**208–215**).[2]

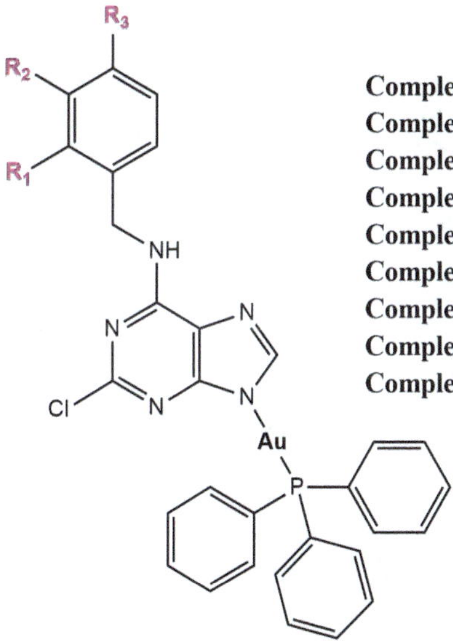

Complex 216: R_1 = H, R_2 = H and R_3 = H
Complex 217: R_1 = H, R_2 = F and R_3 = H
Complex 218: R_1 = Cl, R_2 = H and R_3 = H
Complex 219: R_1 = H, R_2 = Cl and R_3 = H
Complex 220: R_1 = OCH_3, R_2 = H and R_3 = H
Complex 221: R_1 = H, R_2 = OCH_3 and R_3 = H
Complex 222: R_1 = H, R_2 = H and R_3 = OCH_3
Complex 223: R_1 = H, R_2 = H and R_3 = OH
Complex 224: R_1 = H, R_2 = H and R_3 = OCH_3

Figure 8.2 Chemical structures of mixed-ligand gold(I)-triphenylphosphine complexes (**216–224**).[5]

as rheumatoid arthritis. The same research group extended their studies and investigated the anti-inflammatory and anti-edematous activities of mixed-ligand gold(I)-triphenylphosphine complexes (Figure 8.2, complexes **216–224**). The anti-inflammatory activities of the complexes **216, 220** and **224** were investigated using an *in vitro* model based on the expression of pro- and anti-inflammatory cytokines and the influence of the complexes on selected forms of matrix metalloproteinases secreted by LPS-activated THP-1 monocytes. The anti-edematous action of the complexes was evaluated by using an *in vivo* model of carrageenan-induced rat hind paw edema. Low anti-edematous and anti-inflammatory properties of the complexes were reported in both *in vivo* and *ex vivo* investigations, although the *in vitro* studies did indicate promising anti-inflammatory potential of the complexes. The cause of the low activity of the complexes in *in vivo* screening was due to the low stabilities of the complexes in the biological environment, which was evident by the interaction of the complexes with sulfur-bearing biomolecules, cysteine and reduced glutathione.[5] The researchers further extended their studies and documented the *in vitro* and *in vivo* anti-inflammatory properties of trichlorido-gold(III) complexes containing *N*6-benzyladenine derivatives (Figure 8.3, complexes **225–229**). The complexes were investigated for

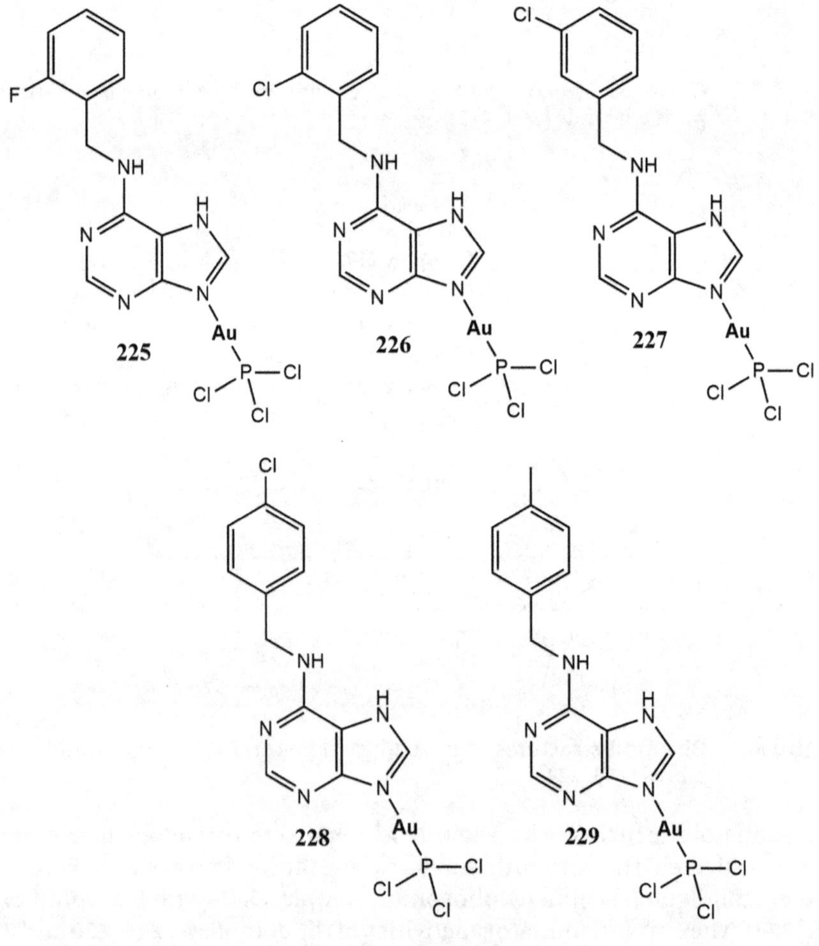

Figure 8.3 Chemical structures of trichlorido-gold(III) complexes containing *N*6-benzyladenine derivatives (**225–229**).[6]

in vitro anti-inflammatory effects towards LPS-activated macrophages (THP-1 cell line), and the *in vivo* studies were carried out by screening the complexes **225**, **226** and **229** against the carrageenan-induced hind paw edema model in rats. The authors were surprised to report a very small *in vitro* anti-inflammatory activity of the complexes, in comparison to a very significant activity *in vivo*. The *in vivo* anti-inflammatory effects of the complexes were comparable with the standard drug indomethacin. In addition, the anti-inflammatory properties of the complexes *in vivo* were even better than for auranofin.[6] The authors continued their investigations on the development of selective anti-inflammatory agents by preparing gold(I) complexes of 9-deazahypoxanthine and triphenylphosphine (Figure 8.4, complexes **230–234**). The complexes were investigated for their *in vitro* anti-inflammatory activities and *in vivo* anti-edematous activities against LPS-activated macrophages and λ-carrageenan-induced hind paw edema rat model. The complexes markedly affected the secretion and expression of TNF-α and IL-1β by a mechanism similar to that of auranofin. In addition, the complexes substantially affected the rate and overall volume of the edema, which was induced by the intraplantar application of λ-carrageenan polysaccharide to rats. The overall results from the research indicated that these complexes are suitable for further screening as candidates for the development of potential anti-inflammatory agents.[7]

Stability of gold complexes under physiological conditions is a major issue in the rational design and development of anti-inflammatory gold complexes. Cyclometalation is recognized as an important approach that increases the stability of gold complexes under conditions of physiological relevance. Cyclometalation helps in restricting the unsought intracellular redox

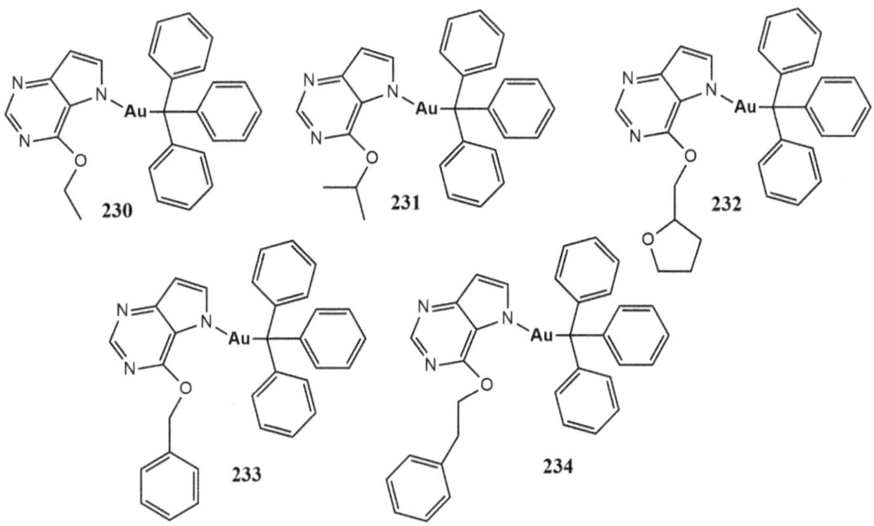

Figure 8.4 Chemical structures of gold(I) complexes of 9-deazahypoxanthine and triphenylphosphine (**230–234**).[7]

reactions of gold complexes due to the incorporation of strong gold–carbon σ bonds.[8] In view of the importance of cyclometalation in the development of stable gold complexes of therapeutic importance, Reddy *et al.*[9] reported the physiological stability, and *in vitro* and *in vivo* anti-inflammatory activities by the study of LPS-activated macrophages and carrageenan-induced hind paw edema in rats, respectively, for four gold(III) complexes containing cyclometalated phosphine sulfide ligands (Figure 8.5, complexes **235–238**). All the complexes were stable under physiological conditions. Complexes **235–238** inhibited the production of TNF-α and IL-1β after LPS treatment. Complexes **236–238** displayed better inhibition of TNF-α expression than that of IL-1β. On the other hand, complex **235** exhibited better IL-1β reduction. Interestingly, the inhibition of the expression of both cytokines by the complexes was similar to that of the standard drugs prednisolone and auranofin. The complexes **235**, **236** and **238** exhibited weak *in vivo* anti-inflammatory activity in the animal models, whereas complex **237** was moderately active, with

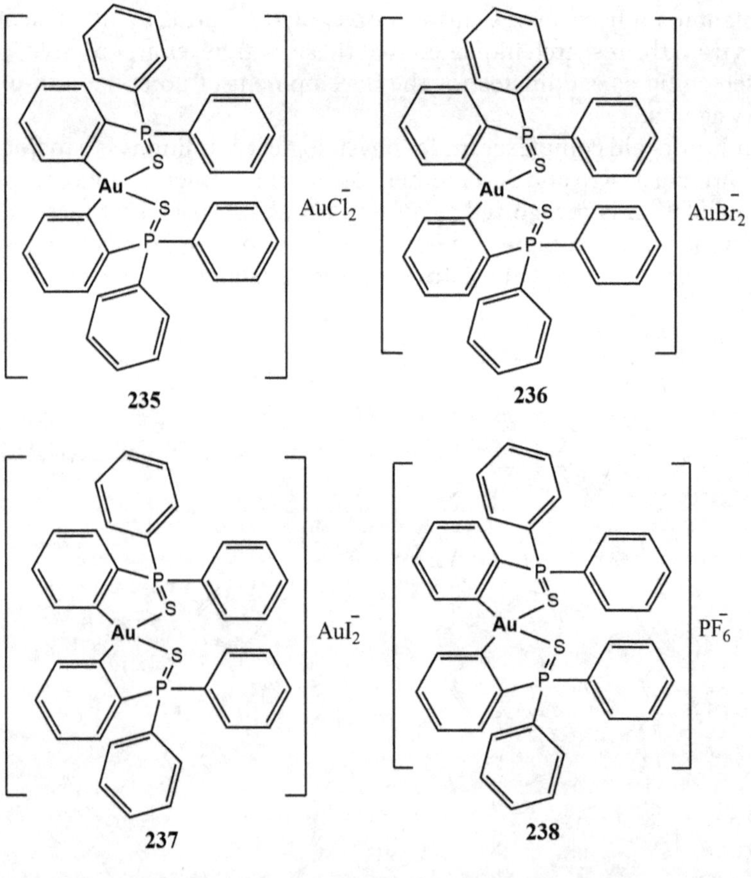

Figure 8.5 Chemical structures of gold(III) complexes containing cyclometalated phosphine sulfide ligands (**235–238**).[9]

an activity comparable to that of auranofin. Overall, the results of the study indicated that cycloaurated phosphine sulfide complexes are interesting candidates, which could serve as important leads for further studies towards the development of safe and effective anti-inflammatory gold complexes.

N-Heterocyclic carbene (NHC) ligand systems represent a very important class of ligands in organometallic chemistry.[10–12] The tunable steric properties of NHCs, brought about by variation in their wingtips, is one of the most important characteristics of NHC ligands.[13] Additionally, the electronic properties of NHCs can be altered by the selection of suitable NHC backbones, in the form of imidazole, triazole, benzimidazole and imidazoline cores, among others.[14–16] Interest in the drug development rationale based on metal complexes of NHC ligands was triggered by the revelation of the bioactivity of Ru(II)-NHC complexes by Çetinkaya and co-workers.[17] Recently, Phung *et al.*[18] documented the anti-inflammatory activities of gold(I) and gold(III) complexes of 1,2,4-triazole-derived carbenes (Figure 8.6, complexes **239–242**). Gold(I) complexes **239** and **240** did not affect the membrane stabilization of red blood cell (RBC) membranes. Gold(III) complexes **241** and **242** potentially inhibited the heat-induced hemolysis of RBCs. The complexes were more active than diclofenac, a well-known non-steroidal anti-inflammatory drug. In addition, the gold(III) complexes **241** and **242** effectively suppressed the generation of NO triggered by LPSs in RAW264.7 macrophages and the complexes displayed very little cytotoxic effect towards normal cells. Interestingly, complexes **241** and **242** showed better inhibitory effects on NO production compared to dexamethasone, a well-known anti-inflammatory drug. Overall, the complexes **241** and **242** showed promising anti-inflammatory properties, and, therefore, their further evaluation in *in vivo* models is warranted. Recently, Trávníček *et al.*[19] reported the *in vitro* anti-inflammatory activities of eight NHC-gold(I) complexes bearing 1,3-bis(2,6-diisopropylphenyl)imidazole-2-ylidene ligands in combination with 6-mercaptopurine

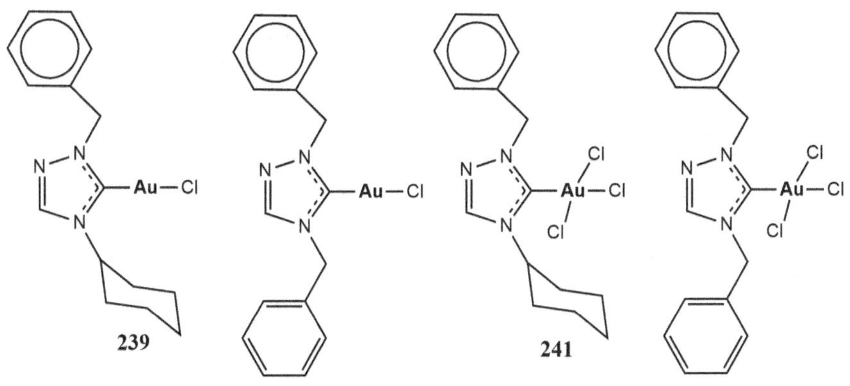

239 **240** **241** **242**

Figure 8.6 Chemical structures of gold(I) and gold(III) complexes of 1,2,4-triazole-derived carbenes (**239–242**).[18]

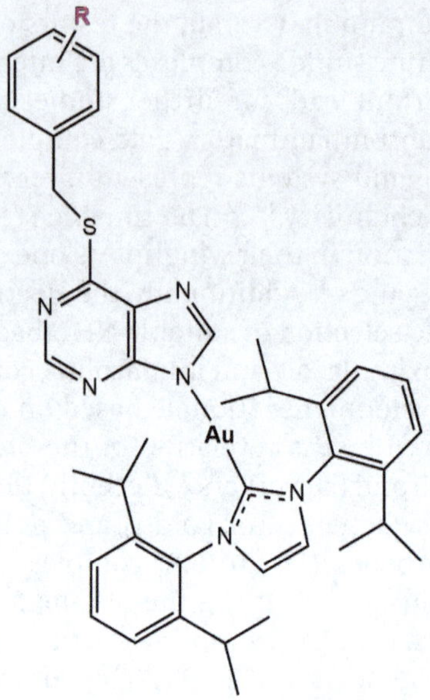

Complex 243: R = *m*-F
Complex 244: R = *m*-CF_3
Complex 245: R = *p*-F
Complex 246: R = *p*-CF_3
Complex 247: R = *o*-NO_2
Complex 248: R = *m*-NO_2
Complex 249: R = *p*-NO_2
Complex 250: R = *p*-Cl

Figure 8.7 Chemical structures of NHC-gold(I) complexes bearing 1,3-bis(2,6-diisopropylphenyl)imidazol-2-ylidene ligands in combination with 6-mercaptopurine (**243–250**).[19]

(Figure 8.7, complexes **243–250**). The complexes negligibly affected the generation of TNF-α and the activation of NF-κB or peroxisome proliferator-activated receptor gamma (PPARγ). The same research group continued their studies and demonstrated the anti-inflammatory activities of gold(I)-NHC complexes bearing a combination of 1,3-bis(2,6-diisopropylphenyl)-imidazole-2-ylidene and the corresponding 7-azaindole derivatives (Figure 8.8, complexes **251–254**). The effects of complexes **251–254** on the activity of NF-κB and secretion of TNF-α were evaluated. The results indicated that complex **254** exhibited comparable activity to that of auranofin.[20]

Generally, inflammatory bowel diseases (IBDs), together with ulcerative colitis and Crohn's disease, are chronic and recurrent inflammatory conditions of the intestinal tract. An alarmingly increasing incidence of IBDs has been reported in the early 21st century in several newly industrialized countries.[21] IBD is a global disease and it seriously risks human health, in addition to putting huge financial burdens on individuals, families and society. Multiple drugs for the treatment of IBDs are available; however, effective drugs without side effects are still unavailable. Krajewska and co-workers[22] reported the *in vitro* anti-inflammatory activities of a water-soluble and photostable gold(III) complex, TGS 121, with chemical formula $[Au(CN)_4]_2(ClO_2)Na$, on a mouse macrophage cell line, RAW264.7. The *in vivo* anti-inflammatory activity of the complex was investigated in the dextran sulfate sodium-induced mouse model of colitis. The complex displayed a potent anti-inflammatory activity in LPS-stimulated RAW264.7 cells, which was indicated by neutral red

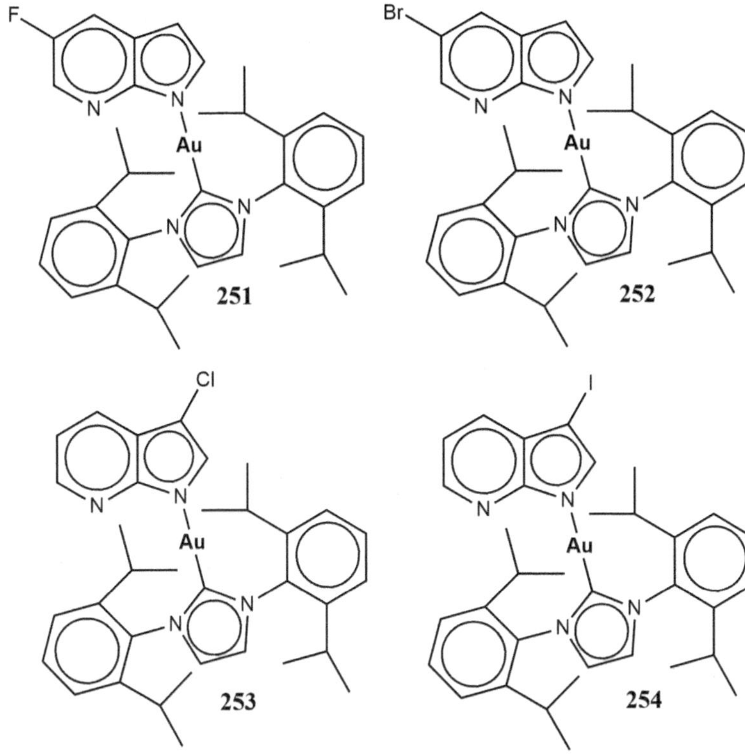

Figure 8.8 Chemical structures of gold(I)-NHC complexes bearing a combination of 1,3-bis(2,6-diisopropylphenyl)-imidazole-2-ylidene and the corresponding 7-azaindole derivatives (**251–254**).[20]

uptake and Griess tests. For investigation in the dextran sulfate sodium-induced mouse model of colitis, the complex was administered in two doses (1.68 $\mu g\ kg^{-1}$, intragastrically, and 16.8 $\mu g\ kg^{-1}$, intragastrically, once daily). The complex displayed a noteworthy anti-inflammatory effect. The anti-inflammatory activity of the complex was attributed to the enzymatic and non-enzymatic antioxidant system. In addition, the complex caused changes in the tight junction complex expression in the intestinal wall. In further studies, a series of gold(III) complexes (TGS 404, 512, 701, 702 and 703) were screened *in vitro*, *in silico* and *in vivo* for inflammation and IBD. In view of the interesting *in vitro* and *in silico* studies, TGS 703, with molecular formula $[Au(CN)_3H_2O]$, was selected for *in vivo* studies. The complex remarkably lessened inflammation in the dextran sulfate sodium-induced mouse model of colitis. The anti-inflammatory activity of the complex was attributed to the enzymatic and non-enzymatic antioxidant systems.[23] An analysis of these two research reports indicates that TGS121 and TGS 703, and other relevant gold(III) complexes, exhibit exciting anti-inflammatory activities and, thus, may be used in treating IBD.

Evaluation of the research findings described in this chapter indicates that several gold(I) and gold(III) complexes with differently structured ligands have displayed exciting anti-inflammatory activities. NHCs, triphenylphosphine ligands, substituted purine derivatives, benzyladenine derivatives, 9-deazahypoxanthine and cyclometalated phosphine sulfide ligands have been employed for the design and development of gold complexes with anti-inflammatory activities. Some of the complexes discussed have strongly reduced the generation of pro-inflammatory cytokines, namely, TNF-α, IL-1β and HMGB1, without influencing production of the anti-inflammatory cytokine, IL-1RA, in LPS-activated macrophages. In addition, some complexes substantially affected edema formation that was triggered by the intraplantar administration of polysaccharide λ-carrageenan to rats *in vivo*, while some complexes effectively suppressed the generation of NO triggered by LPSs in RAW264.7 macrophages. Importantly, several complexes showed comparable and better anti-inflammatory activities than the standard anti-inflammatory drugs, including auranofin, indomethacin, prednisolone, diclofenac and dexamethasone. Overall, the anti-inflammatory gold(I) and gold(III) complexes described here have shown tremendous potential for the rational design and development of future anti-inflammatory agents.

References

1. M. Gielen and E. R. Tiekink, *Metallotherapeutic Drugs and Metal-Based Diagnostic Agents: The Use of Metals in Medicine*, John Wiley & Sons, 2005.
2. Z. Trávníček, P. Štarha, J. Vančo, T. Šilha, J. Hošek, P. Suchý and G. Pražanová, *J. Med. Chem.*, 2012, **55**, 4568–4579.
3. H. J. Williams, J. R. Ward, J. C. Reading, R. H. Brooks, D. O. Clegg, J. L. Skosey, M. H. Weisman, R. F. Willkens, J. Z. Singer, G. S. Alarcón, E. H. Field, P. J. Clements, I. J. Russell, R. F. Hochman, D. T. Boumpas and D. A. Marble, *Arthritis Rheum.*, 1992, **35**, 259–269.
4. E. Alessio, *Bioinorganic Medicinal Chemistry*, John Wiley & Sons, 2011.
5. J. Hošek, J. Vančo, P. Štarha, L. Paráková and Z. Trávníček, *PLoS One*, 2013, **8**, e82441.
6. R. Křikavová, J. Hošek, P. Suchý Jr, J. Vančo and Z. Trávníček, *J. Inorg. Biochem.*, 2014, **134**, 92–99.
7. J. Vančo, J. Gáliková, J. Hošek, Z. Dvořák, L. Paráková and Z. Trávníček, *PLoS One*, 2014, **9**, e109901.
8. T. Zou, C. T. Lum, C.-N. Lok, J.-J. Zhang and C.-M. Che, *Chem. Soc. Rev.*, 2015, **44**, 8786–8801.
9. T. S. Reddy, D. Pooja, S. H. Privér, R. B. Luwor, N. Mirzadeh, S. Ramesan, S. Ramakrishna, S. Karri, M. Kuncha and S. K. Bhargava, *Chem. Eur. J.*, 2019, **25**, 14089–14100.
10. W. A. Herrmann, *Angew. Chem., Int. Ed.*, 2002, **41**, 1290–1309.
11. F. E. Hahn and M. C. Jahnke, *Angew. Chem., Int. Ed.*, 2008, **47**, 3122–3172.
12. S. D. Düşünceli, D. Ayaz, E. Üstün, S. Günal, N. Özdemir, M. Dinçer and İ. Özdemir, *J. Coord. Chem.*, 2020, **73**, 1967–1986.

13. H. Clavier and S. P. Nolan, *Chem. Commun.*, 2010, **46**, 841–861.
14. M. N. Hopkinson, C. Richter, M. Schedler and F. Glorius, *Nature*, 2014, **510**, 485–496.
15. H. V. Huynh, *Chem. Rev.*, 2018, **118**, 9457–9492.
16. S. Guo, J. C. Bernhammer and H. V. Huynh, *Dalton Trans.*, 2015, **44**, 15157–15165.
17. B. Çetinkaya, E. Çetinkaya, H. Küçükbay and R. Durmaz, *Arzneim. Forsch.*, 1996, **46**, 821–823.
18. H. T. Phung, H.-M. Vu, M. Q. H. Ly, H. H. Nguyen, T. H. Nguyen, H. T. Luong and V. H. Nguyen, *Inorg. Chem. Commun.*, 2023, **154**, 110898.
19. Z. Trávníček, J. Vančo, M. Čajan, J. Belza, I. Popa, J. Hošek, R. Lenobel and Z. Dvořák, *Appl. Organomet. Chem.*, 2024, **38**, e7401.
20. J. Belza, Z. Trávníček, J. Vančo, M. Čajan, J. Hošek and Z. Dvořák, *Organometallics*, 2024, **43**, 1155–1164.
21. S. C. Ng, H. Y. Shi, N. Hamidi, F. E. Underwood, W. Tang, E. I. Benchimol, R. Panaccione, S. Ghosh, J. C. Wu and F. K. Chan, *Lancet*, 2017, **390**, 2769–2778.
22. J. B. Krajewska, J. Włodarczyk, D. Jacenik, R. Kordek, P. Taciak, R. Szczepaniak and J. Fichna, *Int. J. Mol. Sci.*, 2021, **22**, 3121.
23. J. Włodarczyk, J. Krajewska, Ł. Szeleszczuk, P. Szałwińska, A. Gurba, S. Lipiec, P. Taciak, R. Szczepaniak, I. Mlynarczuk-Bialy and J. Fichna, *Int. J. Mol. Sci.*, 2023, **24**, 7025.

CHAPTER 9

Nanoformulations of Therapeutic Gold Complexes

Traditional cancer treatments often cause side effects because drugs reach healthy tissues as well as diseased ones. Tumor blood vessels are abnormal and leaky, offering a potential advantage for targeted drug delivery using nanoparticles (NPs).[1] These leaky vessels have pores ranging from 100 nm to several hundred nanometers in size. NPs engineered to this size range could potentially be designed to deliver drugs specifically to tumors (Figure 9.1).[2,3] Leaky tumor blood vessels allow NPs of a similar size to accumulate within tumors, a passive targeting approach that is called the enhanced permeability and retention (EPR) effect. However, poor water solubility and stability limit the use of many drugs. Encasing drugs in NPs can address these issues and improve treatment effectiveness. These carriers can hold high drug payloads and release them steadily.

The limited solubility and chemical stability of Au(I)- and Au(III)-based complexes in the body's natural environment are significant obstacles to their approval by the US FDA for cancer treatment.[4] Additionally, the effectiveness of cytotoxic Au(I)/Au(III) complexes is influenced by their specific form (speciation).[5] However, this speciation often leads to rapid reactions in which biological molecules (nucleophiles) swap places with the bound ligands, especially for Au(III) complexes. This process frequently reduces them to Au(I) or even to metallic gold, Au(0), diminishing their activity and potentially causing side effects.[5] These side effects can arise from the release of either the unbound ligands or the gold itself. Therefore, for gold complexes to be viable cancer treatments, utilizing strategies such as nanoformulation might be essential to improve delivery and overcome these stability issues.

The Therapeutic Potential of Gold Complexes
By Waseem A. Wani, Mohamed F. Alajmi, Afzal Hussain and Irshad Ahmad Wani

Published by the Royal Society of Chemistry, www.rsc.org

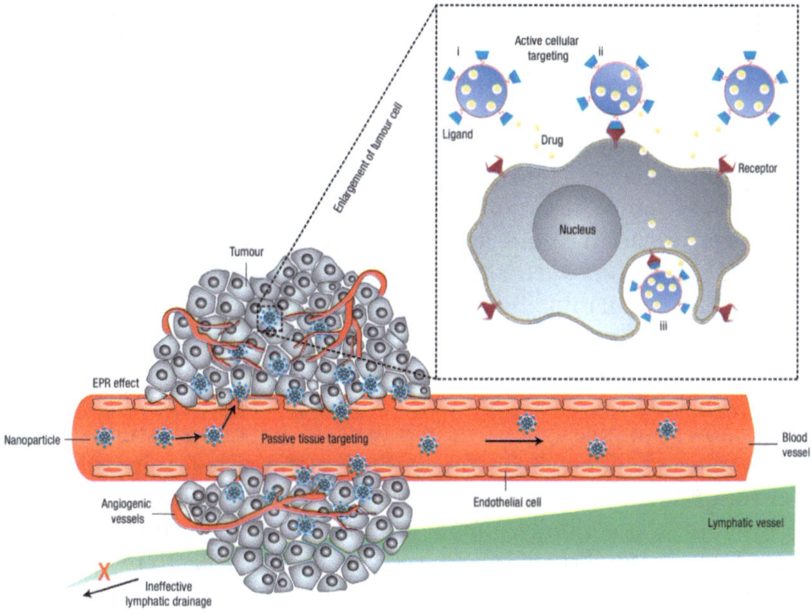

Figure 9.1 Schematic illustration of the two key strategies for delivering drugs to tumors using nanocarriers: passive and active targeting. Passive targeting (represented by circles depicting polymeric nanoparticles) leverages the EPR effect. Leaky tumor blood vessels and poor lymphatic drainage within tumors allow NPs to accumulate there. Active targeting (shown in the inset) involves modifying the nanocarrier surface with ligands (targeting molecules). These ligands bind specifically to receptors on cancer cells, directing NPs towards the tumor and improving drug delivery. Once NPs reach the tumor site, they can: (1) release their drug cargo near the target cells; (2) attach to the cell membrane, acting as a sustained-release depot close to the cells; and (3) enter the cancer cells themselves for intracellular drug delivery. Reproduced from ref. 4 with permission from Nature Publishing Group, Copyright 2007.

The emergence of nanomedicine, which leverages nanotechnology tools for medical purposes, such as drug delivery and imaging using NPs, has revolutionized treatment strategies.[6–8] This field has achieved significant milestones, including approval of the liposomal drug Onpattro for the treatment of polyneuropathy[9] and the development of messenger ribonucleic acid (mRNA) vaccines against SARS-CoV-2.[10] Notably, in nanomedicine, the definition of an NP can extend to particles exceeding 100 nm, as long as they exhibit unique biological or physico-chemical properties absent in the bulk material.[11,12] However, NPs typically have at least one dimension within the 1–100 nm range.[13,14,14–18]

Developing effective gold-based nanomedicines hinges on the strategic use of diverse NPs, while drawing on past experience to optimize the design. Crucially, ensuring biocompatibility of NPs and their components alongside thorough toxicological evaluation is paramount to minimizing potential risks.[19] A recent assessment by the Nanotechnology Characterization Laboratory (NCL), Maryland, USA sheds light on common challenges faced with novel NPs in biomedicine.[19] Building on the limited research on particle-based delivery systems for gold complexes summarized by Lok *et al.* in 2014,[1] significant strides have been made.

Cisplatin resistance in tumors has spurred research into alternative metal-based drugs, including gold and ruthenium complexes. However, these drugs have limitations such as instability, poor water solubility and harsh side effects. Encapsulating them in carriers offers a potential solution. A previously developed stable gold(III)-porphyrin complex (Figure 9.2, compound **255**) effectively combats various cancers, including nasopharyngeal carcinoma and melanoma, demonstrating the promise of this approach.[20] Complex **255** shows promise against various cancers, but animal studies revealed a narrow safety window. To address this, gelatin-acacia microcapsules to deliver complex **255** have been developed.[21] These microcapsules improved water solubility, potentially enhanced tumor targeting and released the drug gradually. Laboratory tests showed sustained cell death in cancer cells. Importantly, mice with tumors treated with microencapsulated complex **255** showed significant tumor suppression with no observed toxicity, unlike mice treated with the free drug at the same dose (75% died). These results suggested microcapsules as a promising strategy to improve the safety and efficacy of complex **255** for cancer treatment. Lee *et al.*[22] developed lipidic gold-porphyrin NPs (GPNPs) using a special technique. These NPs significantly improved the safety profile of complex **255**. In animal studies, a much higher dose of GPNPs (15 mg kg^{-1}) was tolerated compared to free complex **255** (5 mg kg^{-1} caused 50% death rate). Furthermore, GPNPs effectively suppressed tumor growth (86% in neuroblastoma) and extended the life span in mice. These findings highlighted the potential of lipid-based NPs to deliver complex **255** safely and effectively for cancer treatment. Lee *et al.*[23] further improved their GPNPs by adding polyethylene glycol (PEG) to create GPNP-PEG. This modification reduced complex **255** accumulation in liver and lungs, while increasing its concentration in tumor tissue. This suggested that PEGylation enhanced the targeting ability of GPNPs, potentially leading to more effective and safer cancer therapy. These studies demonstrate that micro/nanocarriers effectively reduced the toxicity of free complex **255** and enhance its antitumor effects in animal models.

Supramolecular self-assembly offers a promising alternative to traditional carrier-based drug delivery. This approach relies on non-covalent interactions, such as hydrogen bonding and π–π interactions to create nanostructures. This suggests that self-assembled drug molecules could provide a controlled and sustained drug release mechanism, improving therapeutic efficacy.[24,25] Early research by Davis *et al.*[26] and Yan *et al.*[21] investigated the use of microparticles for delivering gold complexes. They focused on creating microparticles loaded with auranofin (AF), an antitumor agent, using specific materials (monosodium titanate and amorphous peroxo-titanate). These researchers then evaluated the biological effects of these particles in

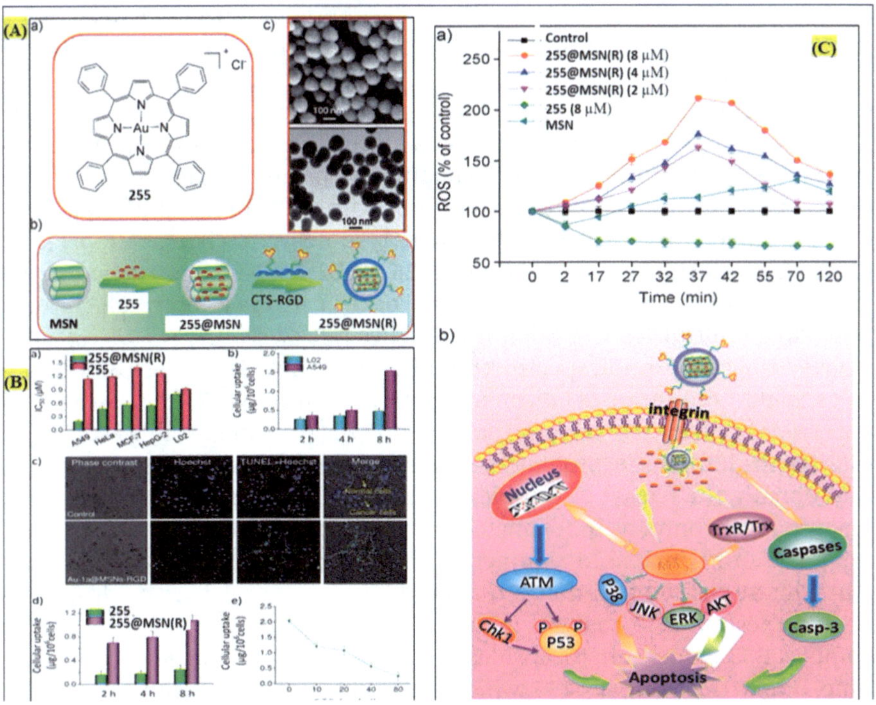

Figure 9.2 (A) (a) The chemical structure of complex **255**. (b) Preparation process of complex **255**-loaded MSPs, labeled as **255**@MSN(R), with RGD represented by a yellow triangle. (c) SEM and TEM images depicting **255**@MSN(R) NPs, scale bars = 100 nm. (B) Selectivity and cellular uptake of **255**@MSN(R). (a) Cytotoxic effects of **255**@MSN(R) and complex **255** on different human cancer and normal cell lines after 72 h of incubation. (b) Quantitative analysis of cellular uptake of **255**@MSN(R) (2.0 mm) in L02 normal human liver cells and A549 lung carcinoma cell lines. (c) Selective induction of apoptosis by **255**@MSN(R) (1.0 mm; 24 h) in an A549/L02 co-culture cell model, as shown by TUNEL-Hoechst 33342 co-staining microscopy (magnification, ×200). (d) Quantitative analysis of cellular uptake in A549 cells after treatment with complex **255** and **255**@MSN(R) (both at 2.0 mm). (e) Dose-dependent effects of RGD peptide on the cellular uptake of **255**@MSN(R). Cells were treated with RGD for 2 h, followed by exposure to **255**@MSN(R) (2.0 mm) for 12 h. (C) Activation of apoptotic signaling pathways in cancer cells by **255**@MSN(R). (a) Graph showing the overproduction of ROS over time in A549 cells exposed to **255**@MSN(R), complex **255** and MSN (1.5 mg mL^{-1}). (b) Proposed signaling pathways of apoptosis induced by **255**@MSN(R). Reproduced from ref. 30 with permission from John Wiley and Sons, Copyright 2014.

cell cultures. Davis[26] explored loading AF onto microparticles made from monosodium titanate and amorphous peroxo-titanate, which they described in detail. Their method achieved only a moderate loading efficiency, which was probably due to limitations of drug solubility in the water-based preparation. These AF-loaded microparticles were then tested in cell cultures.

Compared to free AF, the NPs significantly suppressed succinate dehydrogenase (SDH) activity in fibroblasts, with a dose-dependent effect. At a dose of 5 μg mL^{-1}, the NPs caused significant SDH suppression, and this effect was even stronger at a higher dose of 100 μg mL^{-1} (corresponding to 0.540 μM AF). Interestingly, no significant SDH suppression was observed in monocytes (THP-1) even after 72 h of incubation with the NPs. Yan *et al.*[21] developed a microcapsule delivery system for complex **255**. The microcapsules, made from gelatin, acacia and PEG, were responsive to changes in pH. The researchers encapsulated the complex **255** into the NPs and evaluated the anticancer effectiveness of these particle-based carriers in both cell cultures (including multiple cancer cell lines) and live mice with tumors. Using ultraviolet–visible absorption spectroscopy, they determined that these microcapsules could hold up to 2% of their weight in complex **255**. The researchers observed that these NP-encased gold complexes remained stable even when exposed to glutathione, probably because the NPs formed a protective barrier. Additionally, the release of complex **255** from the NPs was slow and steady in a neutral environment (phosphate-buffered saline at pH 7.0). Interestingly, the release rate increased significantly in acidic environments (pH 5.0 and 3.0). This suggested that these NPs could be useful for delivering drugs in a way that responds to changes in acidity. *In vitro* studies revealed that these complex **255**-loaded microcapsules (**255**@P2) killed HeLa cancer cells over time, unlike free complex **255**, which caused immediate toxicity.

While these microparticle studies showcase the potential of particle-based delivery, there are limitations. Smaller NPs (under 200 nm) are generally considered more favorable for drug delivery. This is because they may trigger weaker immune responses, can be eliminated from the body more easily, release drugs faster and, potentially, even penetrate nearby tissues more effectively.[27–29] In a 2010 study, Lee *et al.* developed a nanometer-sized micellar formulation (around 80 ± 30 nm) for delivering complex **255** to potentially treat neuroblastoma.[22] Direct systematic injections of the nanoformulation into the tumors of mice significantly reduced tumor growth. The dose used was 0.08 mg (roughly 4 mg per kg body weight) every 7 days. Importantly, all treated mice survived the three-week treatment period. This contrasted sharply with mice receiving free complex **255**, where all died within 2 days of treatment (at doses of 4–5 mg kg^{-1}). However, a limitation of encapsulated NPs emerged when high drug doses were needed. At a dose of 0.1 mg, acute toxicity occurred, with only 33% of mice surviving past the second week.[22] The researchers later reported that these micelles were primarily absorbed by liver immune cells (macrophages) after injection into the bloodstream, limiting their effectiveness in cancer treatment. To address this, they modified the micelles with PEG, which reduced liver uptake in mice and increased accumulation of complex **255** within the tumor. However, despite this improvement in targeting, the ability to slow tumor growth was not significantly better compared to the original micelles.[23] Building on the previous work with micelles, He *et al.*[30] explored the use of spherical mesoporous silica nanoparticles

(MSPs) for targeted cancer treatment. These MSPs were roughly 100 nm in size and loaded with the gold complex **255** through an impregnation process (Figure 9.2A). Loading of MSPs with complex **255** caused them to clump together, increasing their size to 589 nm. To improve their targeting ability, a molecule of cyclic tetrasulfide-arginyl-glycyl-aspartic acid (CTS-RGD), which binds to cancer cells, was attached to the surface of the NPs. This reduced the particle size to around 125 nm and generated **255@MSN(R)** NPs, which was confirmed by scanning electron microscopy (SEM), transmission electron microscopy (TEM) and high-resolution TEM (Figure 9.2A). Both **255@MSN** and **255@MSN(R)** NPs had a slightly positive electrical charge. While **255@MSN** NPs by themselves had some anticancer effect, they were also toxic to normal cells. To prevent premature release of the cargo, a biodegradable chitosan shell was subsequently applied around the NPs. To achieve targeted delivery, the chitosan shell was further modified with the tumor-seeking peptide, arginine-glycine-aspartic acid (RGD). Attaching the targeting molecule (CTS-RGD) to the prepared **255@MSN(R)** significantly increased the selectivity of the NPs for cancer cells. Cancer cells treated with **255@MSN(R)** NPs showed a greater increase in the amount of gold taken up by the cells and a higher rate of cell death, compared to normal cells (Figure 9.2B). A loading efficiency of 15% for the gold complex within the particles (150.8 mg g^{-1} of NPs) was achieved. Additionally, efficient cellular uptake of **255@MSN(RGD)** by integrin-expressing cells was also achieved. The chitosan shell facilitated triggered release of the gold complex. Exposure to cellular environments, either through cell lysates or internalization by A549 cells, mimicked the presence of hydrolytic enzymes that degrade the chitosan, promoting complex **255** release. Further investigation revealed that the NPs triggered apoptosis through a specific cellular pathway involving reactive oxygen species (ROS) (Figure 9.2C).

In 2020, researchers proposed a novel approach for systemic delivery of complex **255**.[31] This formulation involved encapsulating the complex within an interpenetrating network made from chemically modified gelatin and polyethylene glycol diacrylate (PEGdA). The researchers proposed that negatively charged parts of the gelatin might interact with the positively charged complex **255** through strong electrostatic forces, leading to slow release of the complex. Despite this, the researchers found that microparticles preserved the ability of complex **255** to kill lung cancer cells (NCI-H460). Animal testing with nude mice implanted with NCI-H460 lung cancer cells (xenografts) revealed similar anticancer effects for microparticles and for free complex **255**.[31] Importantly, microparticles administered intravenously at a dose of 6 mg per kg body weight did not cause any detectable systemic toxicity, unlike free complex **255**, which exhibited significantly higher toxicity. Sun *et al.* investigated a biocompatible and biodegradable zinc metal–organic framework (Zn-MOF) as a delivery system for a powerful anticancer agent (Figure 9.3A, complex **256**).[32] This dinuclear NHC-Au(I)-pyrrolidinedithiocarbamate complex demonstrated significant activity against A2780 cisplatin-resistant cancer cells, with IC_{50} values ranging from 0.83 to 22 μM. Notably,

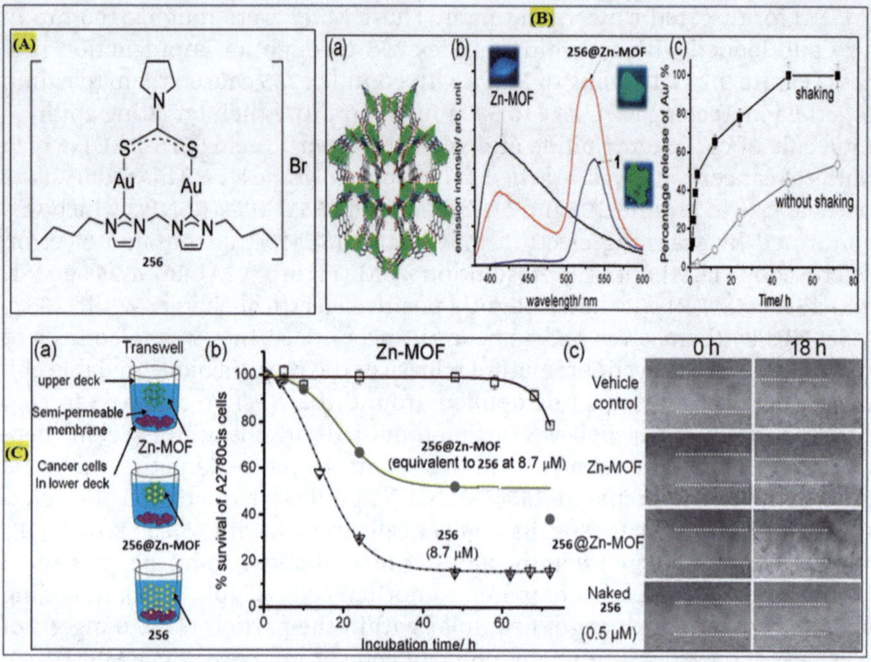

Figure 9.3 (A) Chemical structure of the dinuclear NHC Au(I)-pyrrolidinedithiocarbamato complex **256**. (B) (a) Guest-accessible channels of Zn-MOF observed along the [1,0,1] axis. (b) Photoluminescence spectra of **256**, Zn-MOF and **256**@Zn-MOF (excitation at 364 nm); insets show photographs of the solid samples under a UV lamp with 365 nm excitation. (c) Gold-release profiles of **256**@Zn-MOF with and without shaking at a constant rate of 30 rpm, measured by ICP-AES. (C) Trans-well assay-based studies on cytotoxicity and anti-migration of A2780cis cells. (a) Schematic representation of the trans-well assays. (b) 72 h cytotoxicity profiles of cells with varying co-incubation times with Zn-MOF and **256**@Zn-MOF; time-dependent cytotoxicity of complex **256** at 8.7 μM (naked complex **256**) provided for reference. (c) The impact of Zn-MOF, **256**@Zn-MOF and complex **256** at 0.5 μM on 18 h cell migration. Reproduced from ref. 32 with permission from John Wiley and Sons, Copyright 2015.

complex **256** stood out due to its 40-fold greater toxicity against these cancer cells compared to cisplatin, making it a prime candidate for further development using NPs.[32] The researchers employed a previously reported MOF for this study.[33] This MOF possessed a high void volume (68.5%) and surface area (3057.5 $m^2 g^{-1}$), with appropriate pore size (11 × 8 Å) decorated with hydrophilic groups for better interaction (Figure 9.3B(a)). The MOF particles (10 μm^3) were suitable for intratumoral injection. Successful loading of complex **256** into the MOF was verified through a significant emission change (Figure 9.3B(b)) and confirmed by analytical techniques. The MOF released complex **256** in a controlled manner, reaching nearly complete release (90%)

within 48 h (Figure 9.3B(c)). Complex **256** was incorporated into the MOF using a soaking process. This method achieved a maximum loading capacity of 2.049 mg of complex **256** per gram of MOF.[32] The Zn-MOF carrier acts like a sponge, slowly releasing the complex into surrounding tissues over time (up to 80 h). This sustained release property is particularly attractive for treating cancers resistant to cisplatin, where long-term drug exposure is beneficial. However, directly injecting into the bloodstream (intravenous administration) is not ideal because the particles are too large (over 10 μm). Trans-well assay revealed the cytotoxic and antimigratory effects of complex **256** encapsulated within the Zn-MOF (**256@**Zn-MOF) against A2780cis cancer cells (Figure 9.3D). The MOF functioned as a delivery system, allowing for sustained release of complex **256** and thereby mitigating its initial burst of high cytotoxicity. Notably, the Zn-MOF itself did not exhibit significant cytotoxicity. Both complex **256** and **256@**Zn-MOF effectively suppressed cell migration, suggesting potential for hindering metastasis. Despite the size limitation for intravenous use, both the free complex **256** and the drug-loaded **256@** Zn-MOF demonstrated effectiveness against cisplatin-resistant cancer cells.[32]

In 2015, Lin *et al.* investigated micelle-like NPs for delivering a powerful TrxR inhibitor, complex **257**.[34] These NPs displayed an average size of 50 nm as determined by dynamic light scattering (DLS) and TEM (Figure 9.4A(a and b)), which is suitable for cellular uptake. The researchers employed various techniques, including inductively coupled plasma-atomic emission spectroscopy (ICP-AES), to confirm successful encapsulation of complex **257** within the NPs, with a measured gold content of 2.8% (MLE). The **257@**NPs exhibited significantly higher cytotoxicity against MCF-7 cancer cells compared to free complex **257** or empty NPs alone. The IC_{50} value was 40.0 μg mL^{-1} for **257@**NPs, which is significantly lower than the 1.8 μg mL^{-1} for free complex **257** and 160 μg mL^{-1} for empty NPs (Figure 9.4A(c)). This suggested a synergistic effect between complex **257** and the NPs, potentiating their cell-killing ability (Figure 9.4A(d)). Further investigation revealed that **257@**NPs induced multiple cell death pathways in cancer cells. Flow cytometry analysis demonstrated a higher percentage of apoptotic cells following treatment with **257@**NPs compared to controls. Bio-TEM analysis provided further insights into the cellular uptake and impact of **257@**NPs. The presence of autophagosomes and large vacuoles observed in cells treated with all formulations suggested the induction of autophagy (Figure 9.4B(a)). Notably, **257@**NP treatment resulted in the most significant accumulation of autophagic structures, correlating with increased autophagic cell death. Additionally, western blot analysis revealed that **257@**NPs effectively regulated autophagy (Figure 9.4B(b and c)). Interestingly, the autophagy inhibitor 3-methyladenine (3-MA) further enhanced the cytotoxic effects of **257@**NPs, solidifying the role of autophagy in the cell death process triggered by these NPs. Confocal microscopy confirmed the cellular uptake of NPs through endocytosis, evidenced by the colocalization of Cy5-loaded micelles with endosomes/lysosomes (Figure 9.4C(a)). Furthermore, enzyme

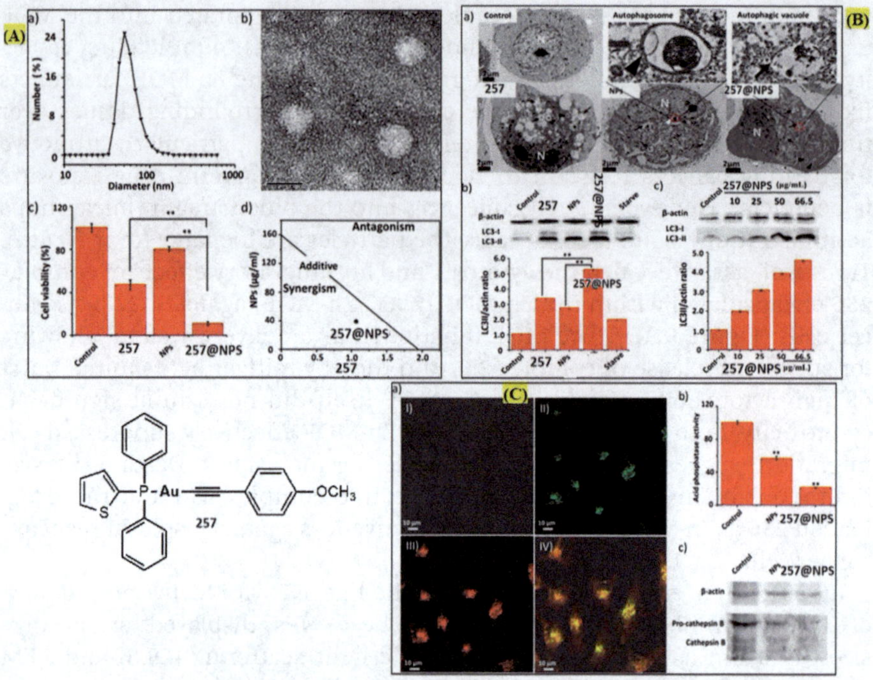

Figure 9.4 (A) Characterization of **257@NPs**. (a) Size distribution of **257@NPs** determined by DLS. (b) TEM images of **257@NPs** with a 50 nm error bar. (c) Viability comparison of MCF-7 cells treated with complex **257** (1.8 μg mL^{-1}), pH-sensitive polymeric NPs (64.7 μg mL^{-1}) and **257@NPs** (66.5 μg mL^{-1}). (d) Isobologram analysis of antiproliferative effects on MCF-7 cells using complex **257**. (B) Synergistic induction of autophagy by pH-sensitive polymeric NPs and complex **257** in MCF-7 cells. (a) TEM images showing untreated (control), complex **257** (1.8 μg mL^{-1}), polymeric NPs (64.7 μg mL^{-1}) and **257@NPs** (66.5 μg mL^{-1}) treatment for 24 h, with labeled nuclei (N) and autophagic structures (arrowheads for autophagosomes and arrows for autophagic vacuoles). (b) Western blot analysis of LC3II expression in MCF-7 cells: untreated (control), starved (Earle's balanced salt solution), and treated with complex **257** (1.8 μg mL^{-1}), polymeric NPs (64.7 μg mL^{-1}) and **257@NPs** (66.5 μg mL^{-1}) for 24 h. Semi-quantitative results shown in the lower panel. (c) Western blot of LC3II expression in MCF-7 cells: untreated (control), starved and treated with varying concentrations of **257@NPs** (10, 25, 50, 66.5 μg mL^{-1}) for 24 h. Semi-quantitative results in the lower panel. (C) Lysosomal impairment by **257@NPs**. (a) Confocal microscopy of MCF-7 cells: (I) blank for cell morphology, (II) green fluorescence for lysosomes stained with LysoTracker Green DND-26, (III) red fluorescence for Cy5-NPs location and (IV) merged image of I, II and III. Cells were incubated with Cy5-NPs (5 μg mL^{-1}) for 2 h. (b) Acid phosphatase enzyme activity in MCF-7 cells after 24 h incubation with NPs (64.7 μg mL^{-1}) and **257@NPs** (66.5 μg mL^{-1}). (c) Western blot analysis of cathepsin B expression in MCF-7 cells after 24 h incubation with NPs (64.7 μg mL^{-1}) and **257@NPs** (66.5 μg mL^{-1}). Reproduced from ref. 34 with permission from American Chemical Society, Copyright 2015.

activity assays and western blot analysis revealed lysosomal damage following **257@NPs** treatment. This damage was characterized by decreased acid phosphatase activity and reduced cathepsin B levels (Figure 9.4C(b and c)). Flow cytometry analysis using LysoTracker Green DND-189 dye provided further evidence of **257@NP**-induced lysosomal dysfunction, indicated by a decrease in lysosomal fluorescence intensity, suggesting lysosome alkalinization. This concept of triggering drug release within specific cellular compartments (endosomal escape) is crucial for designing better drug delivery systems for targeting cancer cells.[35]

Inspired by how ferritin proteins safely store iron within the body, researchers used ferritin cages to deliver the gold complex **258** (Figure 9.5).[36] Apoferritin (empty ferritin) was loaded with complex **258** using a previously established method.[37] This approach leverages the pH-dependent behavior of ferritin, whose subunits can disassemble and reassemble in solution, allowing the gold complex to be trapped inside the ferritin cage.[36] The gold-loaded ferritin cages measured around 13.8 nm with a slight variation (0.9 nm). The researchers investigated the anticancer activity of the caged complex against various human cancer cell lines, including HeLa, MCF-7 and HepG2, and compared it to the non-cancerous cell lines, HaCaT, HRCE and H9c2. While the caged complex exhibited significant cytotoxicity towards all cell lines (IC_{50} in MCF-7 cells was 41 ± 9 μg mL^{-1}), the complex was notably less toxic to non-cancerous cells (IC_{50} in HRCE cells was 61 ± 5 μg mL^{-1}). This

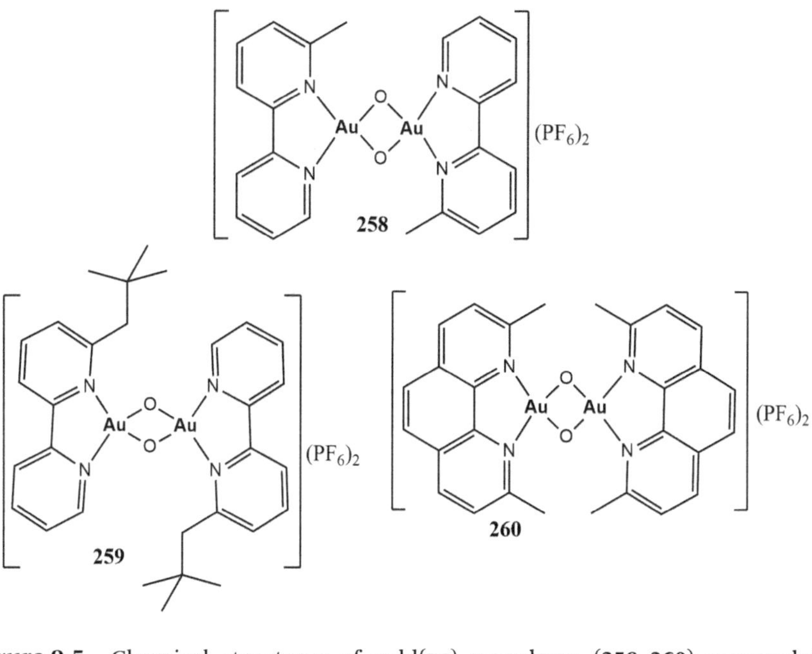

Figure 9.5 Chemical structures of gold(III) complexes (**258**–**260**) encapsulated within ferritin cages.

is a clear contrast to the higher overall toxicity of free complex **258** (IC_{50} in MCF-7 cells was 8 ± 2 μg mL^{-1} and IC_{50} value in HRCE cells was 2.2 ± 0.4 μg mL^{-1}). These results suggested that encapsulation within the ferritin carrier reduced the acute toxicity of the gold complex, possibly due to its partial degradation inside the cage.[38] The same research group also explored encapsulating other gold complexes (Figure 9.5, complexes **259** and **260**) within ferritin cages. Interestingly, these nanoformulations exhibited greater cytotoxicity compared to the previously reported **258**-loaded cages.[39] In a separate study, the authors investigated using ferritin cages for co-delivery of gold and platinum-based drugs:[40] a heterobimetallic compound containing both Pt(II) and Au(I) (Figure 9.6, complex **261**). This approach represented a promising strategy for combination therapy using gold and platinum drugs. However, the encapsulated complex **261** degraded within the ferritin cage. Specifically, Au(I) bound to cysteine 126 within the cage, while Pt(II) fragments were trapped elsewhere inside the carrier. This cargo degradation highlighted a crucial aspect to consider when designing future ferritin-based drug delivery systems. Experiments measuring cell death (cytotoxicity assays) revealed that the free complex **261** was more toxic than its encapsulated form within the ferritin cage.[40] Furthermore, testing across two cancer cell lines (MCF-7 and HeLa) and two non-cancerous cell lines (H9c2 and HaCaT) showed no significant difference in cell death between cancerous and healthy cells. In a recent study, Zhang *et al.* studied the targeted delivery of gold(III) complex **262** (Figure 9.7) by using ferritin-based nanocarriers.[41] The complex exhibited a potent inhibitory effect on glioma cells, with an IC_{50} value of 0.66 ± 0.02 μM. This complex was loaded onto nanocarriers *via* a two-step process. First, disassembly of the ferritin protein was achieved at a low pH (around 2.0). Subsequently, the disassembled ferritin subunits were mixed with the gold complex solution, followed by an increase in pH to 7.5. This final step triggered the reassembly of the ferritin subunits around the encapsulated gold complex (Figure 9.7A(a)). The resulting complex-loaded particles were

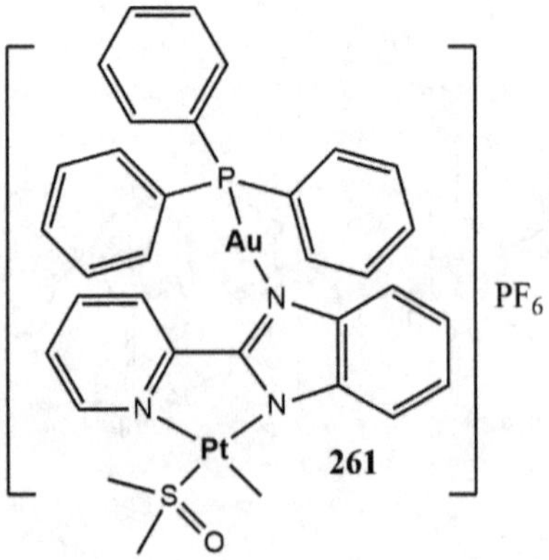

Figure 9.6 Chemical structures of a heterobimetallic complex containing both Pt(II) and Au(I) centers (**261**).

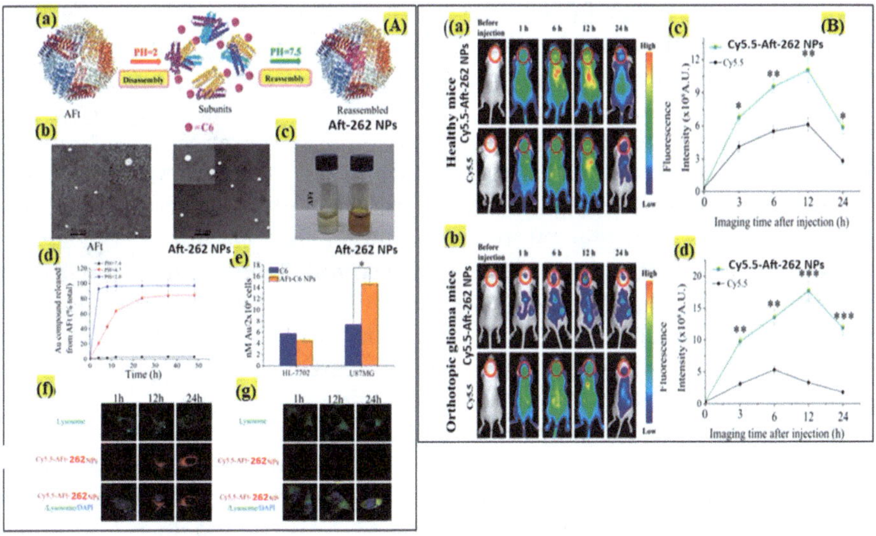

Figure 9.7 (A) Development of the AFt NP-based **262** delivery system. (a) Diagram showing **262** loading into Aft. (b) SEM images of the AFt nanocage (left) and the AFt-**262** NPs (right). (c) Photographs of AFt (left) and AFt-**262** NPs (right) in a glass bottle. (d) *In vitro* release profile of **262** from the AFt-**262** NPs. (e) ICP-MS analysis demonstrating the targeting ability of AFt-**262** NPs to U87MG cells *in vitro*. (f) Confocal images showing intracellular uptake of Cy5.5-labeled AFt-**262** NPs by U87MG tumor cells. (g) Confocal images showing intracellular uptake of Cy5.5-labeled AFt-**262** NPs by HL-7702 cells. $*p < 0.05$. (B) (a) *In vivo* imaging of the brains of healthy mice and (b) tumor-bearing mice intravenously injected with Cy5.5-AFt-**262** NPs. (c) Quantitative analysis of fluorescence signals of Cy5.5-AFt-**262** NPs in the brain areas of healthy mice and (d) tumor-bearing mice. $*p < 0.05$, $**p < 0.02$, $***p < 0.00$. Reproduced from ref. 41 with permission of American Chemical Society, Copyright 2020.

spherical and roughly 13 nm in size [Figure 9.7A(b and c)]. Additionally, their surface charge (zeta potential) was around 18 mV, which is similar to empty ferritin carriers. This similarity suggested that the gold complex was successfully encapsulated within the ferritin cage rather than simply sticking to the outer surface. This approach contrasted with previous studies where the gold complex exchanged molecules with the carrier. Here, the gold complex remained intact within the ferritin carrier. The researchers leveraged the unique pH-dependent behavior of the carrier as it can assemble and disassemble based on the acidity of the environment. While the particles were stable at neutral pH (7.5), they disassembled in acidic environments (around pH 4.6), with a half-life of 10 h. This pH-triggered disassembly facilitated the release of the gold complex with a maximum release observed after 36 h in acidic conditions (Figure 9.7A(d)). The ferritin protein used in the carrier might be responsible for the observed targeted uptake by glioma cells. This is because the transferrin receptor-1 (TfR1) receptor, known to be overexpressed on both blood–brain barrier (BBB) endothelial cells and glioma cells, is recognized by the ferritin protein.[42] Previous research findings suggested TfR1-mediated endocytosis *via* endosomes and lysosomes for apoferritin (AFt) uptake in U87MG cells.[41] In order to validate the proposed mechanism, researchers investigated TfR1 in U87MG and HL-7702 cells using western blotting. They observed higher TfR1 expression on U87MG cells compared to HL-7702 cells (Figure 9.7A(e)). Furthermore, confocal microscopy analysis revealed colocalization of Cy5.5-labeled complex containing NPs with the lysosomal marker lamp1 in U87MG cells (Figure 9.7A(f)), suggesting lysosome-mediated endocytosis as the uptake pathway for AFt-**262** NPs in these cells. To strengthen the endocytosis hypothesis for AFt-**262** NP uptake by glioma cells, U87MG cells were pretreated with chloroquine (CQ), an endocytosis inhibitor, or carbonyl cyanide 3-chlorophenylhydrazone (CCCP), a mitochondrial oxidative phosphorylation uncoupler, before incubation with AFt-**262** NPs. Pretreatment with CQ (20 μM) significantly reduced cellular uptake of AFt NPs in U87MG cells, whereas CCCP pretreatment did not show a substantial effect. These findings collectively point towards endocytosis as the primary mode of AFt-**262** NP uptake by glioma cells. The nanocage successfully crossed the BBB in healthy mice after intravenous injection. *In vitro* cytotoxicity assays against U87MG cells and normal HL-7702 cells revealed that the complex exhibited significantly lower IC_{50} values compared to the corresponding ligands, indicating increased cytotoxicity. These NPs of complex **262** exhibited enhanced cytotoxicity against tumor cells compared to free complex, with minimal side effects, and bioavailability enhancements were also observed. Notably, in mice with glioma tumors, the NP accumulation within the brain region was significantly higher between 6 and 12 h post-injection (Figure 9.7B(a–d)). This targeted delivery approach using ferritin nanocages effectively suppressed glioma tumor growth in mice compared to treatment with either free complex or temozolomide.[41] Mechanistic studies revealed that AFt-**262** NPs induced cell death *via* apoptosis and autophagy pathways, showing mitochondrial dysfunction, ROS generation, endoplasmic reticulum stress and upregulation of apoptotic and autophagy markers.

In a study by Rezaei *et al.*,[43] NPs with a magnetic core and a polymer shell were shown to attract and hold complex **263** (Figure 9.8) through electrostatic

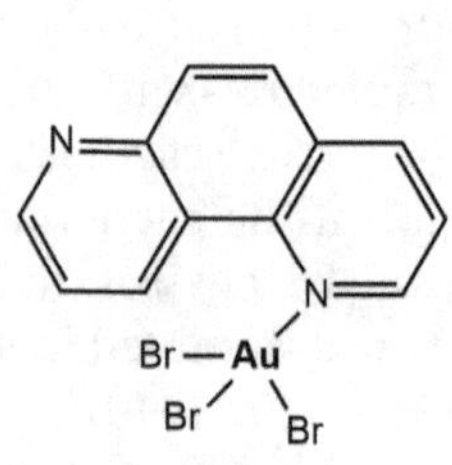

263

Figure 9.8 Chemical structure of anticancer complex Au(1,7-Phen)Br_3 (**263**).

interactions. These 30 nm sized magnetite-core NPs with an acrylic acid/PEG shell were loaded with the gold complex through a soaking process, achieving a capacity of around 9% (confirmed by thermogravimetric analysis and ICP-MS). While the authors suggested a potential for the carboxylate groups to directly bind to the gold center, they concluded that the primary mechanism was electrostatic attraction between the negatively charged carboxylate and the positively charged phenanthroline molecule at physiological pH. The study also revealed that, while drug release was slow at physiological pH, an acidic environment (pH below 5.6) could significantly increase release. Furthermore, in laboratory experiments using HeLa cancer cells, applying an external magnetic field during treatment with the complex-loaded NPs resulted in slightly greater cell death compared to treatment with the free drug alone. This might be because the magnetic field influences the movement of the particles, potentially aiding the entry of these water-soluble carriers into the cells. However, an important limitation of this study is that it did not investigate the stability of the complex in environments mimicking the human body. In a separate study, Malekmohammadi *et al.*[44] explored the use of much larger (around 82 nm) silica particles with mesopores for delivering another gold(III) complex (Figure 9.9, complex **264**) into cells in a pH-dependent manner. The NPs were prepared by attaching targeting molecules like folic acid to the NPs. These folic acid-functionalized NPs were more effective at killing folate receptor-expressing breast cancer cells (MCF-7 line) compared to both regular NPs and the free gold complex they carried.[44]

Attaching metal complexes to nanocarriers can improve their stability, preventing premature release of the drug.[45] However, this approach may require chemical modification of the metal complexes, which can significantly alter the therapeutic activity. For instance, in the development of nanomedicines using gold-based anticancer drugs, two main strategies have been employed. The first strategy involves chemically modifying the gold complexes to make them amphiphilic. These amphiphilic molecules can then self-assemble into nanostructures, encapsulating the drug within. In the second strategy, gold complexes are directly conjugated to biomolecules for transport and protection.[46] Researchers investigated glycopolymer-based micelles as a delivery system for auranofin targeting OVCAR-3 cells.[47] The research team designed block copolymers using reversible addition–fragmentation chain transfer (RAFT) polymerization, which offers precise control over the composition, molecular weight and functionality of the resulting polymers. Three distinct block copolymers, designated

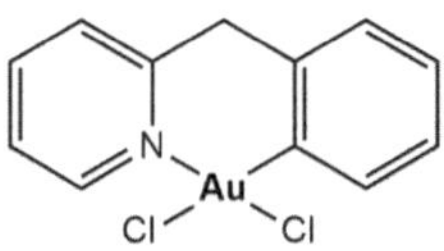

264

Figure 9.9 Chemical structure of a cyclometalated gold(III) complex (**264**).

P20a, P20b and P20c, were synthesized. The efficiency of drug encapsulation (likely an auranofin analogue based on context) within these micelles was assessed. The percentage of drug loaded (complexation yield) was determined using thermogravimetric analysis and found to be 54%, 53% and 63% for P20a, P20b and P20c, respectively. Interestingly, the size of the micelles varied depending on the specific copolymer. P20a self-assembled into the largest micelles at 28 nm, whereas P20b and P20c formed smaller micelles of 23 nm and 9 nm, respectively, as measured by DLS. Researchers investigated the stability of micellar gold complexes using cytotoxicity assays with OVCAR-3 cells. The gold-containing block copolymers (P20a, P20b, P20c) exhibited cytotoxicity in both environments, but to a lesser extent compared to free auranofin. This was confirmed by the IC_{50} values. For example, free auranofin had an IC_{50} of about 1 μM in serum-containing media, whereas P20b and P20c showed IC_{50} values of approximately 16 μM and 19 μM, respectively. Notably, the IC_{50} for P20a was only determined in serum-free media and was 22 μM. These results suggested that the micelles protected the gold complex from being deactivated by biological molecules present in the cellular environment.

A critical assessment of this chapter shows that the use of nanocarriers for delivering gold complexes in cancer treatment is a promising avenue for improved patient outcomes. There is scarce literature available on the nanoformulations of gold complexes for the treatment of bacterial, fungal, viral, parasitic and inflammatory diseases. Therefore, we have restricted our discussion to nanoformulations of gold complexes for cancer treatment only. Non-covalent and covalent attachment strategies have been employed for the delivery of anticancer complexes for the treatment of different types of cancers. Non-covalent strategies offer significant advantages. By encapsulating gold complexes within nanocarriers, such as micelles or ferritin cages, these methods can potentially preserve drug activity while enabling controlled release, improved stability and targeted delivery to cancer cells. This targeted approach can minimize side effects on healthy tissues, a major challenge in conventional cancer therapies. However, challenges remain in optimizing the efficiency of drug encapsulation within the nanocarriers and ensuring effective release of the gold complexes specifically within target cancer cells. Additionally, improving the targeting specificity of these nanocarriers to further minimize uptake by healthy cells is an ongoing area of research. Covalent attachment strategies present an alternative approach. By directly attaching gold complexes to nanocarriers, these methods can potentially improve drug stability during transport within the body. Studies using amphiphilic gold compounds or gold conjugated to biomolecules have shown promising results in terms of cytotoxicity against cancer cells and even targeted delivery with light therapy. However, a key challenge associated with this approach is maintaining the drug's activity after attachment. Additionally, optimizing delivery to diseased cells while minimizing uptake by healthy cells remains crucial.

Looking towards the future, continued research and development of nanocarriers holds immense potential for overcoming these challenges. Some key areas of focus are as follows.

1. *Enhanced targeting capabilities.* Modifying the surface of nanocarriers with specific ligands can improve their ability to recognize and bind to cancer cells, leading to more targeted delivery and reduced off-target effects.
2. *Stimuli-responsive drug release.* Incorporating mechanisms that trigger drug release only in response to specific stimuli within the tumor microenvironment (*e.g.*, pH changes, light exposure, *etc.*) can further enhance treatment efficacy and minimize systemic exposure to the drug.
3. *Combination therapy.* Exploring the synergistic effects of gold complex-based nanomedicines with other agents used in chemotherapy or immunotherapy holds promise for improved treatment outcomes.
4. *Advances in imaging and diagnostics.* These will play a critical role in personalizing cancer therapy. By developing more precise methods for cancer detection and characterization, clinicians can tailor treatment strategies using these nanocarriers to the specific needs of individual patients.

References

1. C. Lok, T. Zou, J. Zhang, I. W. Lin and C. Che, *Adv. Mater.*, 2014, **26**, 5550–5557.
2. D. Peer, J. M. Karp, S. Hong, O. C. Farokhzad, R. Margalit and R. Langer, in *Nano-Enabled Medical Applications*, Jenny Stanford Publishing, 2020.
3. D. Peer, J. M. Karp, S. Hong, O. C. Farokhzad, R. Margalit and R. Langer, *Nat. Nanotechnol.*, 2007, **2**, 751–760.
4. D. Peer, J. M. Karp and S. Hong, *Nat. Nanotechnol.*, 2007, **2**, 751–760.
5. G. Moreno-Alcántar, P. Picchetti and A. Casini, *Angew. Chem.*, 2023, **135**, e202218000.
6. A. Casini, A. Guerri, C. Gabbiani and L. Messori, *J. Inorg. Biochem.*, 2008, **102**, 995–1006.
7. C. L. Ventola, *Pharm. Ther.*, 2017, **42**, 742.
8. K. Riehemann, S. W. Schneider, T. A. Luger, B. Godin, M. Ferrari and H. Fuchs, *Angew. Chem., Int. Ed.*, 2009, **48**, 872–897.
9. A. Akinc, M. A. Maier, M. Manoharan, K. Fitzgerald, M. Jayaraman, S. Barros, S. Ansell, X. Du, M. J. Hope and T. D. Madden, *Nat. Nanotechnol.*, 2019, **14**, 1084–1087.
10. Y. H. Chung, V. Beiss, S. N. Fiering and N. F. Steinmetz, *ACS Nano*, 2020, **14**, 12522–12537.
11. M. Vert, Y. Doi, K.-H. Hellwich, M. Hess, P. Hodge, P. Kubisa, M. Rinaudo and F. Schué, *Pure Appl. Chem.*, 2012, **84**, 377–410.

12. V. Wagner, A. Dullaart, A.-K. Bock and A. Zweck, *Nat. Biotechnol.*, 2006, **24**, 1211–1217.
13. A. Irshad, N. Sarwar and H. Sadia, *Appl. Nanosci.*, 2020, **10**, 3205–3214.
14. T. Ahmad, I. A. Wani, N. Manzoor, J. Ahmed and A. M. Asiri, *Colloids Surf., B*, 2013, **107**, 227–234.
15. I. A. Wani, S. Khatoon, A. Ganguly, J. Ahmed and T. Ahmad, *Colloids Surf., B*, 2013, **101**, 243–250.
16. I. A. Wani, S. Khatoon, A. Ganguly, J. Ahmed, A. K. Ganguli and T. Ahmad, *Mater. Res. Bull.*, 2010, **45**, 1033–1038.
17. I. A. Wani, A. Ganguly, J. Ahmed and T. Ahmad, *Mater. Lett.*, 2011, **65**, 520–522.
18. T. Ahmad, I. A. Wani, I. H. Lone, A. Ganguly, N. Manzoor, A. Ahmad, J. Ahmed and A. S. Al-Shihri, *Mater. Res. Bull.*, 2013, **48**, 12–20.
19. R. M. Crist, J. H. Grossman, A. K. Patri, S. T. Stern, M. A. Dobrovolskaia, P. P. Adiseshaiah, J. D. Clogston and S. E. McNeil, *Integr. Biol.*, 2013, **5**, 66–73.
20. C.-M. Che and R. W.-Y. Sun, *Chem. Commun.*, 2011, **47**, 9554–9560.
21. J. J. Yan, R. W.-Y. Sun, P. Wu, M. C. Lin, A. S.-C. Chan and C.-M. Che, *Dalton Trans.*, 2010, **39**, 7700–7705.
22. P. Lee, Y. Zhu, J. J. Yan, R. W. Sun, W. Hao, X. Liu, C.-M. Che and K. K. Wong, *Nanotechnol., Sci. Appl.*, 2010, **3**, 23–28.
23. P. Lee, R. Zhang, V. Li, X. Liu, R. W. Y. Sun, C. M. Che and K. K. Y. Wong, *Int. J. Nanomed.*, 2012, 731.
24. M. Theodoropoulou and G. K. Stalla, *Front. Neuroendocrinol.*, 2013, **34**, 228–252.
25. C. Valéry, M. Paternostre, B. Robert, T. Gulik-Krzywicki, T. Narayanan, J.-C. Dedieu, G. Keller, M.-L. Torres, R. Cherif-Cheikh, P. Calvo and F. Artzner, *Proc. Natl. Acad. Sci. U. S. A.*, 2003, **100**, 10258–10262.
26. R. R. Davis, D. T. Hobbs, R. Khashaba, P. Sehkar, F. N. Seta, R. L. W. Messer, J. B. Lewis and J. C. Wataha, *J. Biomed. Mater. Res., Part A*, 2010, **93A**, 864–869.
27. A. E. Nel, L. Mädler, D. Velegol, T. Xia, E. M. V. Hoek, P. Somasundaran, F. Klaessig, V. Castranova and M. Thompson, *Nat. Mater.*, 2009, **8**, 543–557.
28. K. Szuhai, A.-M. Cleton-Jansen, P. C. Hogendoorn and J. V. Bovée, *Cancer Genet.*, 2012, **205**, 193–204.
29. S. A. Kulkarni and S.-S. Feng, *Pharm. Res.*, 2013, **30**, 2512–2522.
30. L. He, T. Chen, Y. You, H. Hu, W. Zheng, W. Kwong, T. Zou and C. Che, *Angew. Chem.*, 2014, **126**, 12740–12744.
31. P.-Y. Lee, C.-N. Lok, C.-M. Che and W. J. Kao, *Pharm. Res.*, 2020, **37**, 220.
32. R. W. Sun, M. Zhang, D. Li, Z. Zhang, H. Cai, M. Li, Y. Xian, S. W. Ng and A. S. Wong, *Chem. – Eur. J.*, 2015, **21**, 18534–18538.
33. H. Cai, M. Li, X. Lin, W. Chen, G. Chen, X. Huang and D. Li, *Angew. Chem.*, 2015, **127**, 10600–10605.
34. Y.-X. Lin, Y.-J. Gao, Y. Wang, Z.-Y. Qiao, G. Fan, S.-L. Qiao, R.-X. Zhang, L. Wang and H. Wang, *Mol. Pharm.*, 2015, **12**, 2869–2878.

35. S. A. Smith, L. I. Selby, A. P. R. Johnston and G. K. Such, *Bioconjugate Chem.*, 2019, **30**, 263–272.
36. G. Ferraro, D. M. Monti, A. Amoresano, N. Pontillo, G. Petruk, F. Pane, M. A. Cinellu and A. Merlino, *Chem. Commun.*, 2016, **52**, 9518–9521.
37. X.-T. Ji, L. Huang and H.-Q. Huang, *J. Proteomics*, 2012, **75**, 3145–3157.
38. M. Truffi, L. Fiandra, L. Sorrentino, M. Monieri, F. Corsi and S. Mazzucchelli, *Pharmacol. Res.*, 2016, **107**, 57–65.
39. D. M. Monti, G. Ferraro, G. Petruk, L. Maiore, F. Pane, A. Amoresano, M. A. Cinellu and A. Merlino, *Dalton Trans.*, 2017, **46**, 15354–15362.
40. G. Ferraro, G. Petruk, L. Maiore, F. Pane, A. Amoresano, M. A. Cinellu, D. M. Monti and A. Merlino, *Int. J. Biol. Macromol.*, 2018, **115**, 1116–1121.
41. J. Zhang, Z. Zhang, M. Jiang, S. Li, H. Yuan, H. Sun, F. Yang and H. Liang, *J. Med. Chem.*, 2020, **63**, 13695–13708.
42. T. Moos and E. H. Morgan, *Cell. Mol. Neurobiol.*, 2000, **20**, 77–95.
43. S. J. T. Rezaei, K. Norouzi, A. Hesami, A. M. Malekzadeh, A. Ramazani, V. Amani and R. Ahmadi, *Appl. Organomet. Chem.*, 2018, **32**, e4303.
44. S. Malekmohammadi, H. Hadadzadeh and Z. Amirghofran, *J. Mol. Liq.*, 2018, **265**, 797–806.
45. C. Li, J. Wang, Y. Wang, H. Gao, G. Wei, Y. Huang, H. Yu, Y. Gan, Y. Wang and L. Mei, *Acta Pharm. Sin. B*, 2019, **9**, 1145–1162.
46. S. Shen, Y. Wu, Y. Liu and D. Wu, *Int. J. Nanomed.*, 2017, **12**, 4085–4109.
47. S. Pearson, H. Lu and M. H. Stenzel, *Macromolecules*, 2015, **48**, 1065–1076.

CHAPTER 10

Toxicity of Gold Complexes

Recent studies suggest a possible link between gold(III) formed from gold(I) drugs and the toxic side effects observed during chrysotherapy. In mice chronically treated with gold(I) drugs, T cells were activated by gold(III), not gold(I).[1,2] Similarly, patients with gold-induced dermatitis exhibited significant lymphocyte proliferation in response to gold(III) but not to gold(I).[3] Furthermore, researchers investigated the reaction of bovine ribonuclease (RNase) A, a model antigen, with gold(III). This reaction led to the sensitization of T cells in mice to novel "cryptic peptides" derived from the protein. In contrast, T cells from mice injected only with RNase became sensitive solely to a well-known, immunodominant peptide from RNase.[1,4,5]

A new study proposes that the fragmentation and presentation of hidden peptide fragments (cryptic peptides) might result from two mechanisms triggered by gold(III).

1. Gold(III)-induced oxidation of methionine residues within the protein.
2. Gold(III)-induced conformational changes in the protein, leading to altered intracellular processing and the formation of different antigenic peptides.

The key finding here is that the gold atom itself does not directly participate in the recognition process between the major histocompatibility complex (MHC) class II protein, peptide and T cell. Instead, it influences an earlier step where gold(III) modifies the protein, which then undergoes different processing within the cell.

Gold(I) requires strong oxidizing agents to convert to gold(III). However, these oxidants might be present within the body during inflammation. For instance, the enzyme myeloperoxidase, found in phagocytic cells, can oxidize gold(I) aurothiomalate to gold(III) in the presence of hydrogen peroxide and chloride ions. This occurs *via* the production of hypochlorite (ClO^-).[1,6]

The Therapeutic Potential of Gold Complexes
By Waseem A. Wani, Mohamed F. Alajmi, Afzal Hussain and Irshad Ahmad Wani

Published by the Royal Society of Chemistry, www.rsc.org

Hypochlorite is synthesized by myeloperoxidase during the oxidative burst, a process where these cells generate ROS. Notably, hypochlorite can also oxidize gold(I) in other drugs, such as aurothioglucose and auranofin.[1]

Recent research has prompted us to study how gold(III) interacts with peptides. There is a clear bias in gold research, with most studies focusing on gold(I) and its interactions with naturally occurring thiols. Although gold(III) is used as a heavy atom label in protein X-ray crystallography, its biological chemistry remains relatively unexplored.[7] Only a few investigations have examined how gold(III) reacts with amino acids and peptides, and these have primarily focused on sulfur-containing amino acids, like cysteine and methionine. Interestingly, gold(III) can oxidize disulfide bridges in albumin.[8] The reaction between tetrachloroaurate and cysteine (represented as RSH) is described by the following equation.[9]

$$\mathrm{NaAuCl_4 + 3RSH \rightarrow RSAu + RSSR + 3HCl + NaCl}$$

Similarly, cystine (with two disulfide bonds) can reduce gold(III) to form metallic gold.[10]

$$\mathrm{3RSSR + 10KAuBr_4 + 18H_2O \rightarrow 6RSO3^- + 10Au^0 + 40Br^- + 10K^+ + 36H^+}$$

This reaction likely involves breaking the disulfide bond and further oxidizing the resulting fragments.

Studies show that insulin, a protein with two chains linked by disulfide bridges, can be oxidized by gold(III) compounds (tetrabromoaurate in this case).[8] This oxidation disrupts the insulin structure, potentially affecting its function. This ability of gold(III) to damage protein structure through disulfide bond oxidation might contribute to the toxicity observed with gold(III) complexes, and thus limit their use.[10] Gold(III) can also oxidize methionine and methionine residues in proteins, with these reactions being rapid and specific.[11,12] Interestingly, gold(III) treatment of ribonuclease causes aggregation when the protein is folded, but oxidation of methionine residues to methionine sulfoxide when unfolded.[13] Little is known about gold(III) interacting with other peptide amino acids. Only two crystal structures of gold(III)-peptide complexes exist, both involving the dipeptide glycyl-L-histidine (Gly-L-His).[14] These structures highlight the different binding modes of gold(III) depending on the pH. A four-coordinate gold(III) complex with a tripeptide (Gly-Gly-L-His) under acidic conditions (pH 1.5) has been synthesized recently.[15] Gold(III) stands out for its ability to significantly increase the acidity of a specific nitrogen atom (pyrrole nitrogen) within the histidine imidazole ring when bound to it. This effect, called "across-the-ring ionization", makes this nitrogen atom more likely to lose a proton (become deprotonated). The pK_a value of this nitrogen drops from above 14, in free histidine, to around 9 when bound to gold(III).[16] This influence on acidity is stronger than with most other metal ions. For instance, copper(II) and palladium(II) only decrease the pK_a value to around 11, while ruthenium(III), cobalt(III) and mercury(II) achieve similar reductions to gold(III), but not as dramatically.

These findings suggest that gold(III) can undergo redox reactions with peptides and proteins, especially those containing sulfur-based amino acids. However, gold(III) also demonstrates a unique ability among metal ions, *e.g.* deprotonating and binding to peptide amide bonds and cross-linking histidine imidazole rings. While interactions with other amino acid side chains, such as tyrosine, seem possible, they have not been explored so far. Further research on how gold(III) interacts with peptides and proteins could improve our understanding of the immune system response to gold-based drugs, particularly considering the potential role of gold(III) in the side effects observed during gold therapy.

While gold-based therapeutics offer promise for rheumatoid arthritis, inflammatory diseases and autoimmune conditions, potential organ system toxicity cannot be ignored. Notably, the most severe gold toxicity is typically confined to skin and mucous membranes, with reports arising from uncontrolled blind clinical trials where dose adjustments were unavailable.[17] However, Ciftci *et al.*[18] investigated the potential cardiotoxic effects of gold complexes in rats. Their findings revealed significant oxidative damage within heart tissues. The study showed a dose-dependent increase in malondialdehyde levels, a marker of lipid peroxidation and cellular damage, upon exposure to the gold complexes. This effect mirrored the oxidative damage observed with cisplatin. Additionally, the gold complexes caused a decrease in antioxidant enzyme activity, including superoxide dismutase, catalase and glutathione, highlighting their ability to induce oxidative stress. Notably, this suppression of antioxidant activity was more pronounced with increasing gold complex concentrations. In comparison to ruthenium complexes, the gold complexes induced a greater degree of oxidative stress and cardiotoxicity. This suggested that gold might be a safer alternative to ruthenium for clinical applications in cancer treatment.

The previous study by Gottlieb *et al.*[19] examined two cases of vasomotor (nitritoid) reactions following the administration of gold sodium thiomalate (GST) in rheumatoid arthritis patients and its potential to cause acute myocardial infarction. In the first case, a 69 year-old man experienced a severe vasomotor reaction and subsequently developed an acute anterior wall myocardial infarction, while the second case involved a 49 year-old man who was monitored during GST administration and exhibited sinus tachycardia and a drop in blood pressure, indicating a vasomotor reaction. These reactions involved peripheral vasodilatation due to arteriolar smooth muscle relaxation, similar to nitrite effects, which can compromise myocardial perfusion and increase myocardial work and oxygen demand due to adrenergic discharge. The study suggested that aurothioglucose, which rarely causes vasomotor reactions, may be a safer alternative to GST for older rheumatoid arthritis patients with cardiovascular disease or atherosclerosis. Given the higher incidence of vasomotor reactions with GST and the potential for serious cardiovascular events, the study recommended cautious use and close monitoring of gold complexes in rheumatoid arthritis therapy, particularly favoring aurothioglucose in high-risk patients. The study by Ahmed *et al.*[20] investigated the renal and hepatic toxicity of a newly synthesized gold(III) complex (Figure 10.1, complex **265**) in rats. It employed a two-pronged approach, evaluating both acute and subacute toxicity effects. High doses (1500 and 750 mg kg^{-1}) of the gold(III) complex proved detrimental, leading to significant mortality in rats before tissue collection. Examination of surviving animals revealed varying degrees of renal tubular necrosis (grades 1–5) across these groups, highlighting dose-dependent kidney damage (Figure 10.1). Interestingly, lower doses (375 and 187.5 mg kg^{-1}) still resulted in severe renal damage in many animals, indicating a narrow window of safe dosage in the acute setting. However, a glimmer of hope emerged at the lowest dose (93.75 mg kg^{-1}), where less extensive necrosis was observed. This suggested a potential safe zone for acute administration, but further exploration is needed. Beyond the kidneys, the gold(III) complex also exerted effects on the liver. Findings included variable extents of micro- and macro-vesicular steatosis (fatty build-up), congestion and occasional issues with blood flow within the liver (sinusoidal obstruction syndrome). As with the kidneys, the severity of these changes increased with higher doses. The subacute toxicity study offered a more optimistic outlook. When administered at a dose of 32.2 mg kg^{-1} (one-tenth of the lethal dose), the gold(III) complex displayed minimal negative effects. Kidney histology appeared normal with only mild inflammation and congestion. Similarly, liver changes were limited to mild to moderate ballooning degeneration (cellular swelling) with minimal inflammation. A crucial comparison is drawn with existing chemotherapeutic agents, such as cisplatin, which is notorious for causing significant kidney damage even at lower doses. This gold(III) complex presented a potentially safer profile, demonstrating a broader range of safe doses for the kidneys and a notably higher maximum safe dosage. These findings suggested that complex **265** may offer a less nephrotoxic alternative for cancer treatment.

Janković *et al.*[21] reported the toxicity of several gold complexes (Figure 10.2, complexes **266**, **277** and aurothiomalate) on the spontaneous motility of isolated human fallopian tubes. The results showed that complex **267** has a concentration-dependent increase in the stimulation of spontaneous contractions of the isolated fallopian tube isthmus, while the other complexes do not affect the spontaneous contractions of the isolated fallopian tube ampulla. The research also discussed the potential adverse effects of gold complexes on fertility, citing studies that have shown that gold complexes can inhibit sperm motility, decrease Leydig cell secretory function and decrease prostatic secretory function, leading to male infertility.[22,23] Additionally, the authors also noted that gold complexes can also interfere with the motility of hollow organs, such as the fallopian tubes, which could potentially affect fertility.

In conclusion, the toxicological aspects of gold complexes have been explored in this chapter, with particular emphasis on their importance in the design of safer and more effective gold-based therapies. This chapter highlights the need for further investigation into the interactions between gold(III) complexes and peptides/proteins. Understanding these interactions

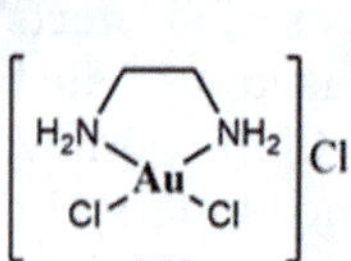

265

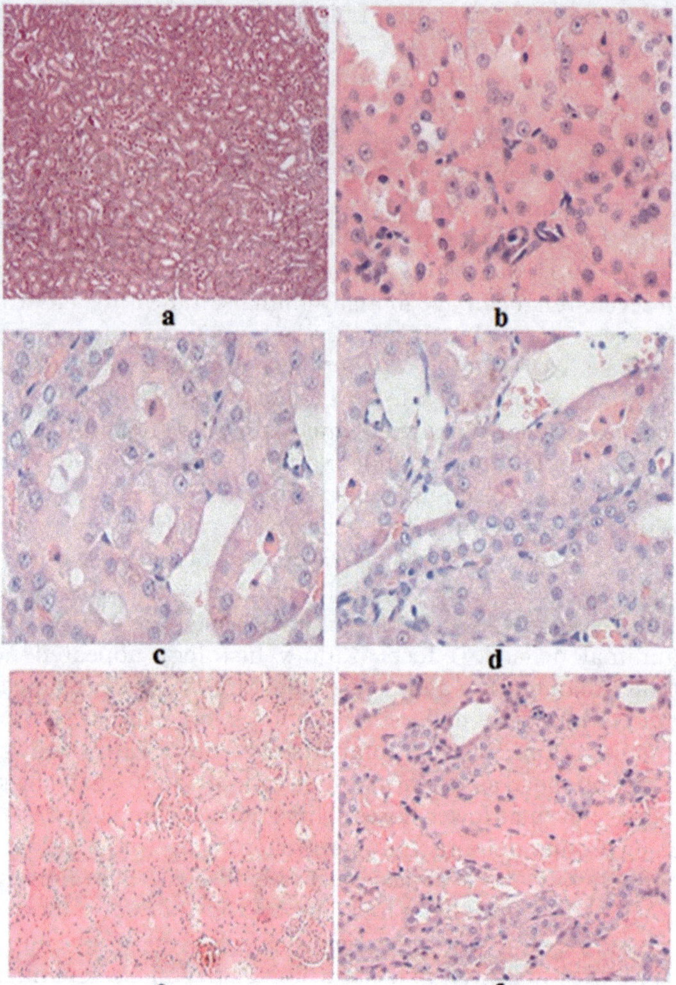

Figure 10.1 Microscopic observations of renal tubules displaying various grades of necrosis from the acute toxicity of complex **265**. (a and b) Grade 2 necrosis under H&E staining at ×620 and ×640 magnification, with necrotic tubules among viable ones, comprising less than 25% of the examined material. At ×640, there are more numerous necrotic cells alongside normal tubules. (c and d) Grade 1 necrosis at ×640 magnification, showing scattered apoptotic/necrotic cells with strongly eosinophilic cytoplasm and pyknotic nuclei. (e and f) Grade 5 necrosis at ×620 and ×640 magnification, where the entire field is predominantly filled with necrotic renal tubules. Reproduced from ref. 20, https://doi.org/10.1371/journal.pone.0051889, under the terms of the CC BY 4.0 license, https://creativecommons.org/licenses/by/4.0/.

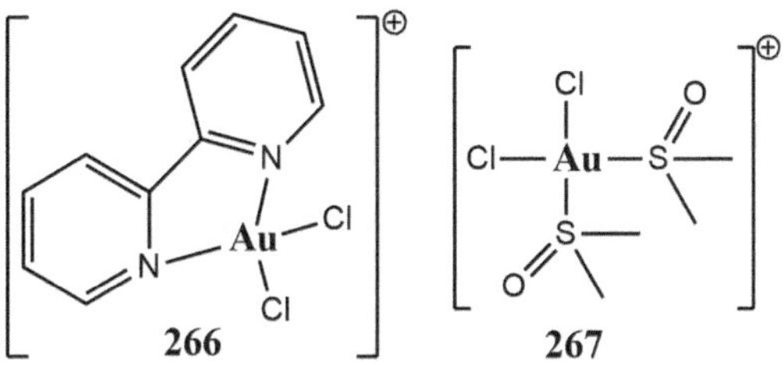

266 **267**

Figure 10.2 Chemical structures of some gold(III) complexes. Adapted from ref. 21 with permission from Springer Nature, Copyright 2012.

is considered crucial for the development of safer and more effective gold-based therapies. The potential mechanisms by which gold(III) might exert its toxic effects have also been discussed. These mechanisms include protein structure modification and histidine residue deprotonation. The suggestion is made that the unique ability of gold(III) to modify protein structures and deprotonate histidine residues offers potential explanations for its toxicity. By unraveling these mechanisms, researchers would be able to pave the way for the next generation of gold-based therapies with minimal side effects. Gold-based therapies hold therapeutic promise, but their potential for organ toxicity, notably in the heart, kidneys and liver, necessitates careful consideration. The studies discussed highlight the importance of monitoring and dose management, especially in high-risk populations, such as older patients with cardiovascular or renal issues. While newer gold complexes show potential for reduced toxicity compared to traditional gold compounds, further research is essential to fully understand their safety profiles and optimize their clinical use. Long-term studies and investigations into other potential toxicities are crucial for ensuring the safe and effective use of these compounds in medical practice.

References

1. S. L. Best and P. J. Sadler, *Gold Bull.*, 1996, **29**, 87–93.
2. D. Schuhmann, M. Kubicka-Muranyi, J. Mirtschewa, J. Günther, P. Kind and E. Gleichmann, *J. Immunol.*, 1990, **145**, 2132–2139.
3. J. Verwilghen, G. H. Kingsley, L. Gambling and G. S. Panayi, *Arthritis Rheum.*, 1992, **35**, 1413–1418.
4. K. M. Pollard, D. K. Lee, C. A. Casiano, M. Bluthner, M. M. Johnston and E. M. Tan, *J. Immunol.*, 1997, **158**, 3521–3528.
5. P. Griem, K. Panthel, H. Kalbacher and E. Gleichmann, *Eur. J. Immunol.*, 1996, **26**, 279–287.
6. B. Beverly and D. Couri, in *Federation Proceedings*, FEDERATION AMER SOC EXP BIOL 9650 ROCKVILLE PIKE, BETHESDA, MD 20814-3998, 1987, vol. 46, pp. 854–854.

7. L. N. Johnson, in *New Comprehensive Biochemistry*, ed. A. Neuberger, and L. L. M. Van Deenen, Elsevier, 1985, vol. 11, pp. 347–415.
8. P. L. Witkiewicz and C. F. Shaw, *J. Chem. Soc., Chem. Commun.*, 1981, 1111–1114.
9. W. E. Smith and J. Reglinski, *Perspect. Bioinorg. Chem.*, 1991, **1**, 183–208.
10. C. F. Shaw, M. P. Cancro, P. L. Witkiewicz and J. E. Eldridge, *Inorg. Chem.*, 1980, **19**, 3198–3201.
11. E. Bordignon, L. Cattalini, G. Natile and A. Scatturin, *J. Chem. Soc., Chem. Commun.*, 1973, 878–879.
12. G. Natile, E. Bordignon and L. Cattalini, *Inorg. Chem.*, 1976, **15**, 246–248.
13. A. A. Isab and P. J. Sadler, *Biochim. Biophys. Acta, Protein Struct.*, 1977, **492**, 322–330.
14. M. Wienken, B. Lippert, E. Zangrando and L. Randaccio, *Inorg. Chem.*, 1992, **31**, 1983–1985.
15. S. L. Best, T. K. Chattopadhyay, M. I. Djuran, R. A. Palmer, P. J. Sadler, I. Sóvágó and K. Varnagy, *J. Chem. Soc., Dalton Trans.*, 1997, 2587–2596.
16. R. J. Sundberg and R. B. Martin, *Chem. Rev.*, 1974, **74**, 471–517.
17. R. Rau, *Clin. Rheumatol.*, 2005, **24**, 189–202.
18. O. Ciftci, I. Ozdemir, O. Cakir and S. Demir, *Toxicol. Ind. Health*, 2011, **27**, 735–741.
19. N. L. Gottlieb and E. H. Brown Jr, *Arthritis Rheum.*, 1977, **20**, 1026–1028.
20. A. Ahmed, D. M. A. Tamimi, A. A. Isab, A. M. M. Alkhawajah and M. A. Shawarby, *PLoS One*, 2012, 7, e51889.
21. S. M. Janković, A. Djeković, Ž. Bugarčić, S. V. Janković, G. Lukić, M. Folić and D. Čanović, *BioMetals*, 2012, **25**, 919–925.
22. S. Suzuki, N. Yamauchi, T. Miyamoto and M. Muranaka, *J. Allergy Clin. Immunol.*, 1983, **72**, 469–474.
23. D. K. Ledford, Y. Sakamoto, C. O'Connor, K. Krzanowski, R. F. Lockey, J. B. Polson and J. Krzanowski, *Arch. Int. Pharmacodyn. Ther.*, 1989, **298**, 237–246.

CHAPTER 11

Future Research Outlook

This book highlights the recent advances in the design and development of therapeutic gold complexes for the treatment of cancer, bacterial, fungal, viral and parasitic diseases, and inflammatory conditions. The discussion in Chapter 3 reveals that a good number of gold complexes based on different ligand architectures with interesting anticancer properties have been reported. The complexes have shown excellent anticancer activities towards several cancer cell lines, such as A2780R, MCF-7, SH-SH5Y, HTC116, HEP3B, A549, SKOV-3, A375, HEK-293, HeLa, PC3, HT1080, MDA-MB-231, K562, H460, OVCAR8, MDA-MB-468, 4T1 and BT-333 cell lines, among others. In addition, several gold complexes with different ligand frameworks have exhibited promising activity and selectivity against different pathogenic bacterial strains, such as MRSA, *Enterococcus faecium*, *Escherichia coli*, *Klebsiella pneumonia*, *Pseudomonas aeruginosa*, *Acinetobacter baumannii*, *Bacillus subtilis* and *Salmonella typhi*, fungi, such as *Candida albicans*, *C. tropicalis*, *Aspergillus flavus*, *C. krusei*, *C. glabrata*, *C. parapsilosis*, *C. auris* and *Cryptococcus neoformans*, viruses, such as HIV, coronavirus, Chikungunya virus and human cytomegalovirus, and parasites, such as *Plasmodium berghei*, *Trypanosoma brucei*, *T. cruzi*, *Leishmania infantum*, *L. major*, *L. amazonensis*, *L. braziliensis*, *L. donovani* and *Toxoplasma gondii*. Additionally, several complexes have displayed promising activities against several inflammatory conditions, including rheumatoid arthritis and inflammatory bowel disease.

Despite the interesting therapeutic profiles of several gold complexes, many other gold complexes have shown insignificant therapeutic profiles. An analysis of the literature revealed that the rational development of therapeutic gold complexes started from the 1890s following the exploration of the bacteriostatic activities of gold cyanide. However, investigations of the therapeutic properties of gold complexes have increased significantly during the last couple of decades. Although several important leads have been

The Therapeutic Potential of Gold Complexes
By Waseem A. Wani, Mohamed F. Alajmi, Afzal Hussain and Irshad Ahmad Wani

Published by the Royal Society of Chemistry, www.rsc.org

obtained so far, with some entering clinical practice, there are still impediments to the design and development of safe and effective therapeutic gold complexes.

For developing effective therapeutic gold complexes, it is very important to thoroughly study the mechanisms of the bioactivities of the active gold complexes. It is evident from the literature that for a few active gold complexes their DNA-binding ability correlates with their anticancer properties. However, several active gold complexes have not shown any noticeable affinity for DNA interactions, although the complexes have displayed their anticancer activities through generation of ROS, inhibition of TxrR, G1 cell cycle arrest, activation of endoplasmic reticulum stress and inhibition of the redox antioxidant transcription factor NRF2. It is important to note that the mechanisms of antibacterial, antifungal, antiviral, antiparasitic and anti-inflammatory activities of gold complexes have been studied, but further studies are needed to determine the facts with regard to the quantitative structure activity relationships of the complexes. Overall, an improved understanding of the cellular effects of bioactive gold complexes might offer future guidelines in the rational design and development of potential therapeutic leads.

The stability of therapeutically active gold complexes under physiological conditions is a massive challenge for their development as effective therapeutics. The physico-chemical characteristics of the ligands in gold complexes determine their pharmacokinetic properties. In comparison to several therapeutically active platinum complexes, certain gold(I) and gold(III) complexes possess improved selectivity and activity. The therapeutic advantages of gold complexes are derived from several complex interactions between the gold center, cellular components and genes.

The biological activity of gold complexes can be enhanced by altering their delivery mechanisms. Nanotechnological approaches make it possible that metallodrugs with target-specific action, and consequently fewer side/adverse effects, can be developed by synthesizing their nanoparticulate forms. Therefore, functionalizing or encapsulating therapeutically active gold complexes into appropriate nanostructures can help to enhance their bioavailability, selectivity, site-specificity and speed of action.[1] In addition, different types of drug delivery systems can be employed for transporting therapeutically active gold complexes to their target sites. It would be interesting to evaluate the activities of newly synthesized gold complexes on their target sites *via* drug delivery systems that are decorated with cell-specific antibodies. The utilization of drug combination approaches can also increase the efficiency of less active therapeutic gold complexes. Since drug combination approaches have shown synergistic effects in a number of biological studies, and, therefore, drug combination studies of gold complexes with therapeutically potent agents may be conducted for arriving at new drug combinations for treating several diseases. At present, a number of software and simulation programs, along with theoretical analysis protocols, including AutoDock, Percepta Platform and Lipinski's rule, are employed for estimation of the therapeutic potency of drugs, even before their synthesis. Thus, molecularly screened gold complexes with promising bioavailability, solubility, stability in physiological conditions, fewer side/adverse effects and high effectiveness can be rationally designed and developed. In addition, a number of computer-based simulation programs are being used for the identification of the biological targets of drugs. Thus, such programs must be used for the design and optimization of the structural features of new gold complexes. Often, several complex biochemical pathways are involved in cancer and in certain diseases of bacterial, fungal, viral, parasitic and inflammatory origin. Thus, the successful treatment of these diseases requires pharmaceutical intervention in multiple pathways, and often through a combination of different drugs. Designed multiple ligands (DMLs) act simultaneously on multiple biological targets, and thus the development of gold complex-based DMLs can be used as a novel strategy to combat diseases effectively. However, so far, only a few gold complexes are used in the treatment of several diseases, but we are confident that the design and development of therapeutically active gold complexes will continue to grow as their therapeutic potential has been revealed through preliminary investigations. We look forward to more efforts from global research groups that will ensure the entry of potential gold complexes into clinic settings.[2] Keeping in view exciting examples of gold complexes with promising therapeutic profiles, further research in the design and development of active gold complexes is warranted, and may eventually pave the way for the discovery of gold complexes with distinct action mechanisms, in comparison to traditional metallodrugs, for the effective treatment and management of cancer and other diseases.

References

1. I. Ali, K. Salim, M. A. Rather, W. A. Wani and A. Haque, *Curr. Cancer Drug Targets*, 2011, **11**, 135–146.
2. C.-H. Leung, H.-J. Zhong, D. S.-H. Chan and D.-L. Ma, *Coord. Chem. Rev.*, 2013, **257**, 1764–1776.

Subject Index